Introduction to
PLANT BREEDING

The Author

Fredrick Mugendi Njoka started his long journey of education in 1979 at Nthima Primary School in Meru District, Kenya. He later joined Nairobi School in 1986 for his Secondary Education where he excelled and joined the Peoples' Friendship University of Russia (PFUR) in 1991. He obtained his BSc. Honours in Agriculture (1996), MSc. Honours in Agriculture (1998) and PhD in Plant Breeding and Seed Production (2001) from PFUR. Mugendi was first employed by the Ministry of Agriculture as a Seed Inspector in 2002. His main activities included variety release, export clearance for farm produce, enforcing adherence to biosafety regulations among others. He joined Kenyatta University in 2003 as a lecturer and rose through the ranks to the post of Dean of Student Affairs in Ruiru Campus. In 2013, Mugendi joined University of Embu (UoEm) as a Senior Lecturer as well as the founding Dean, School of Agriculture. He has actively taken part in student training and research supervised more than twenty masters students and three PhD students. He has published more than 20 articles in refereed journals and participated in local as well as international conferences. His main areas of interest are plant breeding, genetics and general agriculture.

Basing on Mugendi's wealth of experience in teaching, research and knowledge dissemination, he has authored the book "*Introduction to Plant Breeding*" to benefit students in universities and those in institutions of higher learning.

Introduction to
PLANT BREEDING

– The Chief Editor –

Prof. Olieg Grigorievich

Peoples' Friendship University of Russia

Kruger Brentt
P u b l i s h e r s
2 0 1 8

Kruger Brentt Publishers UK. LTD.
Company Number 9728962

Regd. Office: 68 St Margarets Road, Edgware, Middlesex HA8 9UU, UK

Library of Congress Cataloging-in-Publication Data

Cataloging in Publication Data--DK
Courtesy: D.K. Agencies (P) Ltd. <docinfo@dkagencies.com>

Njoka, Fredrick M. (Fredrick Mugendi), author.
Introduction to plant breeding / the author, Fredrick M. Njoka.
        pages cm
        Includes index.
        ISBN 978-1-78715-008-9 (Hardbound)

        1. Plant breeding.   I. Title.

LCC SB123.N56 2017 | DDC 631.52   23

For information on all our publications visit our website at http://krugerbrentt.com/

Preface

A need for sufficient and better food during the development of human civilization has stimulated an ancient human endeavor to start selecting, caring, saving and re-growing the best plant types or better food production, which resulted in domestication of many "selected" wild plant types. This casual selection was changed to direct efforts to retain plants with the most distinct, superior and desirable traits, as improved cultivars to produce better food. To respond to the increasing need to feed the world's population as well as an ever greater demand for a balanced and healthy diet, there is a continuing need to produce improved new cultivars or varieties of plants, particularly crop plants. The strategies used to produce these are increasingly based on our knowledge of relevant science, particularly genetics, but involves a multidisciplinary understanding that optimizes the approaches undertaken.

The book is an introductory course in plant breeding and can be used as supplementary text for both graduate and diploma students. The following features give details of the ways in which the book is organised:

Introductions: Gives a general overview of every topic.

Self-assessment questions (SAQ): These are designed to help you think about what you are reading as you go along. If you do not understand the question, make a note of it and discuss it with your lecturer or do further reading.

Notes: These are designed to help you make emphasis on specific contents of a particular section. They will provide a useful collection of revision material.

Self tests: There are one or more self tests in some chapters. These will help you identify what you have not understood or remembered from that particular chapter.

Past examination questions: These are included to help you get an idea of how this course is examined. The questions are included at the end of chapter ten. The questions have been selected to improve your knowledge of that topic and also to give you some practice in attempting examination questions.

Summary: Summaries are given at the end of every chapter. They will help you discover what you were expected to have covered within the chapter. They will provide a useful collection of revision material.

Further readings: A list of books will be provided after each lesson, indicating text books and other resources that provide materials covered in the chapter. These are important to further proceed your reading.

Introduction to Plant Breeding has been written with the intention of providing introductory knowledge on Plant breeding with emphasis on practical applications. The book endeavors to make this seemingly difficult subject understandable and enjoyable for both students in universities and other related institutions. It provides an account of modern techniques in plant breeding. The book has employed simple learning methods such as self tests, quizzes, illustrations and sample questions.

Fredrick M. Njoka

Contents

Chapter 1

Introduction to Plant Breeding

1.0 Overview

Plant breeding is a branch of science that deals with changing the genetic constitution of a plant to suit human needs both agronomically and economically. Plant breeding can be accomplished through many different techniques ranging from simply selecting plants with desirable characteristics for propagation, to more complex molecular techniques. Scientists discovered that environment could not increase the yield alone. The limit of a yield is set by the genetic make up of a plant. You can give a plant enough water, nutrients, protect it from diseases but if its genetic make up is poor, you will not get a good yield.

1.1 History of Plant Breeding

Just like anything we see has a history, plant breeding has its history as well. Plant breeding was not created by God but man had to do something about it. Plant breeding started with what is called *domestication* - process of bringing wild species under human management. Man went to the fields or forest and chose certain plants for cultivation. Domestication continues up to today and may continue in future especially in timber trees and medicinal plants.

One example of domestication in recent years is the use of *Penicillium* to get an antibiotic called penicillin. Another example of domestication is the transfer of genes for disease resistance from wild species to cultivated ones.

When man moved from one place to another he carried his cultivated species with him. This process of taking a genotype or a group of genotypes from one geographical area to a new one where they were not grown before is called *introduction*.

1.2 Examples of Early Plant Breeders

☆ 700 B.C.– Babylonians and Assyrians pollinated date palm artificially.

☆ 1717 – Thomas Fairchild produced the first artificial hybrids.

☆ 1760 – 1766 – A German scientist by name Joseph Koelreuter crossed different tobacco species (*Nicotiana rustica* x *N. paniculata*) and got hybrids. He became the first man to talk about advantages of interspecies hybridization and the use of heterosis.

Heterosis (hybrid vigor) is the advantage of an offspring over its parents in the first filial generation (F_1).

☆ 1759 – 1835 An English man T. Knight crossed different species of peas and made new varieties.

☆ 1856 – L. Vilmorin used progeny test to improve sugar beets (*Beta vulgaris*).

☆ 1900 – The science of genetics was begun when Gregor John Mendel's papers were discovered. They had been published in 1866.

☆ 1903 – A Danish biologist studied variation in beans for seed weight and proposed the pure line theory necessary for individual selection.

☆ 1905 – Biffen demonstrated that resistance to yellow rust is dominant over susceptibility.

1.3 Disciplines Involved in Plant Breeding

A discipline cannot stand alone and succeed. That's why it is important to go through some major disciplines that are connected to plant breeding.

Note: The basis of plant breeding is genetics and cytogenetics.

Combined efforts of specialists is very important in plant breeding.

1. Genetics and cytogenesis - A breeder must know genetic principles and laws in order to effectively breed a plant.

2. Botany -A Breeder must be well versed with the morphology, reproduction and taxonomy of plants. To deal with a disease or pest you have to be able to classify the plants, area of attack *etc.*

3. Agronomy - To deal with a disease you must consider the way a crop is grown, its health and its capability to resist diseases.

4. Plant physiology - A Breeder studies the functioning of a plant especially the effects of insect and disease attacks.

5. Plant pathology - Knowledge of plant diseases and pathogens especially when breeding for resistances is very important.

6. Entomology - In order to create resistant crops to insects, it is necessary to consider the damage caused and the insect behaviour.

7. Biometry - It is necessary for comparison of performance and correct interpretation of acquired results. Experimental layout is based on biometry.

8. General knowledge - A Breeder must have general knowledge on economics, market demands and predict the future results.

1.4 Activities in Plant Breeding

The desired changes in genotypes are brought about by the following activities: Creation of variation, selection, evaluation and distribution.

Note: Without variation, there can be no improvement of a species.

1. Creation of variation or use of existing one.

    This can be natural or artificial as shown below

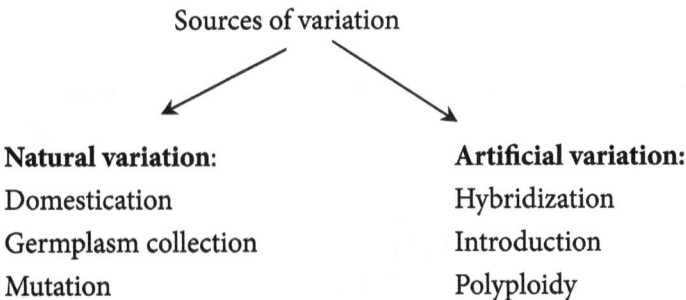

Sources of variation

Natural variation:	**Artificial variation:**
Domestication	Hybridization
Germplasm collection	Introduction
Mutation	Polyploidy

2. Selection – isolation of specific genotypes from a mixed population.

3. Evaluation - comparison of newly acquired genotypes with the existing ones usually for several years.

4. Distribution - if a newly acquired variety has benefits over existing one, it is multiplied and distributed.

1.5 Main Objectives of Plant Breeding Today

Objectives of plant breeding vary from one crop to another depending on its economic significance.

1. Yield improvement - A new variety can only be multiplied and distributed if its yield exceeds that of existing varieties. The yield must be stable.

2. Quality improvement - Quality is a complex trait that is specific to given a crop.

For example:

- ☆ Wheat – Grain size, gluten content, milling and baking qualities, 1000 kernel weight.

- ☆ Barley - Can be for making beer or fodder. Barley for beer usually has big grains, higher germination percentage, low protein, high malting quality.

- ☆ Fruits and vegetables - size, colour, taste, vitamins, sugar content, transportability and storage.

- ☆ Legumes and cereals – Protein content especially the type of amino acid (lysine, methionine, tryptophan).

- ☆ Potatoes - Those used for planting should have many buds and of medium size.

Note: Quality has a negative correlation with many agronomic factors. Examples:

1. Big orange is usually seedless, has poor storage and doesn't propagate easily.

2. Maize pioneer has big grains but is attacked by weevil (Osama).

3. Mangoes that are budded are large but cracked and have poor seed.

4. Quick maturity is associated with poor taste.

1.6 Factors to Consider while Choosing a Plant Breeding Method

1. Ecological plasticity - The ability of a plant to give high and stable yield under climatic stress.

2. Suitability for mechanization - a good variety must be good for mechanization during planting, weeding, fertilizer application, harvesting *etc.* It must be resistant to lodging. In maize the position of the cob from the ground should not be less than 30cm. In sunflower the florescence should be at the same angle and should have minimum shattering.

3. Synchronous (uniform) maturity period. This is very important in legumes *e.g.* garden peas, beans. If maturity period is different, then we need several picking periods.

4. Dormancy. – Best varieties are those with a longer dormancy period. This is because they do not germinate before harvesting. The biggest problem is found in barley, and wheat. They form *volunteer crops*. Volunteer crops appear spontaneously from the past crop without planned planted.

5. Elimination of toxic substances *e.g.* Erusic acid in brassica, Tannin in sorghum and Alanine in potatoes.

6. Pest and disease resistance. This is applicable to all crops. Usually pest and disease outbreak is provoked by increased fertilizer application and extreme weather conditions.

7. Shortening the vegetative period of some crops *e.g.* cotton, pigeon peas.

8. Drought resistance.

1.7. Comparison between Plant Breeding for Resistance and the Use of Pesticides

Disadvantages of using Pesticides

1. Pesticides increase the cost of production.

2. Use of pesticides reduces the population of both harmful and useful insects *e.g.* pollinators.

3. Extensive use of pesticides leads to development of resistant pathogens and insect pests.

4. Wide use of pesticides causes both air and soil pollution.

5. Some pesticides lower the quality of products.

6. Some pesticides are harmful for human consumption.

1.8 Undesirable Consequences of Plant Breeding

Just like all activities have their positive and negative sides, the same happens with plant breeding.

1. Cost of producing a variety is very high because a specialist is needed and it takes quite a long time to train.

2. Genetic erosion – Disappearance of various forms of a cultivated species or its wild relatives. To solve this we have gene banks.

3. Reduces variability due to replacement of heterogeneous local varieties by few dominant more improved varieties.

4. Breeding can create abnormal living things and unexpected plant products *e.g.* genetically engineered foods, bringing up a plant that causes drought.

Summary

In this chapter we have discussed how plant breeding started. We have also highlighted objectives and activities in plant breeding. Consequences of plant breeding have been discussed.

Further Reading

1. B. P. Singh (1995). Plant Breeding. Kalyani Publishers, 677 pages.

2. V. L. Chopra (1989). Plant Breeding: Theory and Practice. Oxford Publishers, 471 pages.

3. A. Dafni. (2001). Field methods in pollination ecology. University of Haifa, Israel. P. 17-30.

4. J. M. Poehlman (1959). Breeding Field crops. University of Missouri, 427 pages.

5. R. W. Allard (1960). Principles of plant breeding. John Wiley and Sons, Inc. California. 485 pages.

6. D. J. Van de have (1979). Heterosis in plant breeding, Proceedings of the 7th Congress of EUCARPIA. Budapest. 365 pages.

7. Y. L. Gushov, A. Fuks and P. Valichek (1999). Breeding and seed production of cultivated crops. Moscow. ISBN 5-209-00964-5. 536 pages.

8. F. V. Guliaev and Y.L. Gushov (1987). Breeding and seed production of cultivated crops. Moscow. 447 pages.

Chapter 2

Evolution of Cultivated Crops

2.0 Introduction

Evolution – gradual process of change and development of a species.

Man is known to have first sown seeds using primitive tools about 20,000 years ago. He domesticated plants and selected the best types. He discovered that the offspring from good seeds were better than those from poor seeds.

According to the time they came to cultivation, plants can be divided into 3 main groups:

1. Without a defined or exact time of first cultivation. This represents the oldest crops to be cultivated.

   Barley is the oldest crop to be cultivated. Others include banana, yam, cassava, cotton, onion, grape, arrowroot, sorghum, sweet potato and millet.

2. First cultivation between 4,000 and 8,000 yrs ago: Rice, oat, soybean, sugarcane, garden peas, maize and sunflower.

3. Those recently cultivated (in 19th century): Sugar-beet and rubber. Some crops are being domesticated even today.

2.1 Patterns of Evolution in Crop Plants

Selection by man and nature has been responsible for evolution of crop plants. However, selection is effective in altering a species only when there is variability in a population of that species.

Natural selection (survival for the fittest) - when a genotype that is more adapted to a given environment leaves behind more progeny than a less adapted species, with time the less adapted becomes extinct.

Evolution comes after a long time of variation and selection.

According to the origin or source of variation, evolution of crops can be grouped into 3 patterns:

1. Mendelian variation
2. Interspecific hybridization
3. Polyploidy.

2.1.1 Evolution Pattern by Mendelian Variation

This is evolution caused by gene mutation and hybridization between different genotypes of the same species. Mutation is a heritable change in organisms. Mutation can be harmful or beneficial. Species carrying harmful mutations are eliminated with time. According to their effect, mutation can be grouped into:

(i) Macromutation
(ii) Micromutation

(i) Macromutation

Causes a large distinctive morphological effect and often expressed on several characters.

Example: Maize (*Zea mays*) was once a grassy, pod corn. A single macromutation led to the modern maize with different position of ear and tassels, and several other characters.

Cabbage (*Brassica* family) originated from a common wild species but has differentiated to broccoli, cauliflower and kale (Figure 2.1).

(ii) Micromutation

Causes small and less drastic changes in organisms. They cause the greatest part of variation in crop species. Some crops that originated from Mendelian variation include barley, rice, beans, peas and tomatoes.

2.1.2 Interspecific Hybridisation

This is crossing 2 different species of plants. It is usually accompanied by some difficulties *e.g.* difference in the number of genes, sterile seeds and different flower structures. Today it is made easy by biotechnology. It is common in vegetatively propagated crops *e.g.* roses, oranges, strawberry, pears and avocado.

Broccoli

Cauliflower

Kale

Figure 2.1: Various Species that Differentiated from a Common Cabbage.
(*Source*: Gushov, 1999).

2.1.3 Polyploidy

It is when an individual carries more than 2 sets of chromosomes. Normal organisms have 2n =2x (diploid). Polyploidy can be 3x (triploid), 4x (tetraploid), 5x (pentaploid), 6x (hexaploid) *etc.*

Autopolyploidy appears due to duplication of chromosomes from the same species.

Examples of autopolyploids include:

Commercial banana (*Musa paradisiaca*), Potatoes (*Solanum tuberosum*), Sweet potato (*Ipomea batata*).

Because the chromosomes are of the same type, autopolyploidy has contributed little to crop evolution.

Allopolyploidy – Results from chromosome doubling of F_1 hybrids crossed from different species. It is mainly influenced by man.

Summary

We have learnt the meaning of evolution and various classifications of plants according to the time they came to first cultivation. We have also learnt about patterns of evolution of cultivated crops. The role of autopolyploidy and allopolyploidy in plant breeding has been discussed.

Further Reading

1. Y. L. Gushov, A. Fuks and P. Valichek (1999). Breeding and seed production of cultivated crops. Moscow. ISBN 5-209-00964-5. 536 pages.

2. V. L. Chopra (1989). Plant Breeding: Theory and Practice. Oxford Publishers. 471 pages.

3. A. Dafni. (2001). Field methods in pollination ecology. University of Haifa, Israel. pp. 17-30.

4. J. M. Poehlman (1959). Breeding field crops. University of Missouri, 427 pages.

5. R. W. Allard (1960). Principles of plant breeding. John Wiley and Sons, Inc. California. 485 pages.

6. D. J. Van de have (1979). Heterosis in plant breeding, proceedings of the 7th Congress of EUCARPIA. Budapest. 365 pages. Y.L. Gushov, A. Fuks and P. Valichek (1999). Breeding and seed production of cultivated crops. Moscow. ISBN 5-209-00964-5. 536 pages.

7. F. V. Guliaev and Y.L. Gushov (1987). Breeding and seed production of cultivated crops. Moscow. 447 pages.

Chapter 3

Centre of Genetic Diversity/Origin

3.0 Introduction

These are regions where crop plants are known to have first evolved from wild species. There is good evidence that cultivated plants were and are not distributed uniformly throughout the world.

A Russian Scientist, (N.I. Vavilov, 1920) proposed that crop plants evolved from wild species in the areas showing great diversity.

3.1 Significance of Centres of Origin

In the centers of origin, we find: Wide variety of species, drought, disease and pest resistance and ideal environment of a species.

3.2 Methods of Detecting the Origin of a Species

- *(i)* *Botany:* Identifying where a species grows spontaneously without man's interference.
- *(ii)* *History:* Historians have tried to explain the origin of many species.
- *(iii)* *Archeology and paleontology:* Detection of recognizable fragments in ancient buildings structures and soil. These can be fruits, seeds, plant parts, drawings especially in pyramids.
- *(iv)* *Philology:* Development of languages.
- *(v)* Combination of different methods.

3.3 Types of Centres of Origin

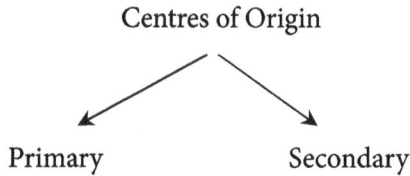

Centres of Origin

Primary   Secondary

A primary centre of origin is a region where a crop plant first evolved from wild species.

A secondary centre of origin is a region where a crop plant did not first evolve but is widely cultivated than in its origin. Example: Coffee originated in Ethiopia but it is cultivated more in Brazil. This means that Brazil is the secondary centre of origin while Ethiopia is the primary centre of origin for coffee. Secondary centres of origin generally lack the richness in diversity found in the primary centre though there can be exceptions.

N. I. Vavilov proposed the law of homologous series in variation which states: *"Characters in one species also occur in other related species"*. This means that a character missing in one species can be found in a related species.

3.4 World Centres of Origin

Based on germplasm data collection, Vavilov proposed 8 centres of origin of crop plants and subsidiary centres. A few variations to Vavilov's discoveries have been recorded. The following are the 8 centres proposed by N. I. Vavilov (Figure 3.1):

Abyssinia, China, Hindustan, Central Asia, Asia Minor, Mediterranean, Central and South America.

1. Abyssinian centre of origin (Ethiopia and Eritrea).

   Barley (*Hordeum vulgare*), Durum wheat (*Triticum durum*), Emmer wheat (*Triticum dicoccum*), Sorghum (*Sorghum bicolor*), Castor (*Ricinus communis*), Coffee (*Coffea arabica*), Onion (*Allium cepa*), Sem (*Dolichos lablab*). Secondary centre of origin of Broad bean (*Vicia faba*).

2. China centre of origin (it is the largest and oldest centre of origin).

   Soybean (*Glycine max.*), Orange (*Citrus sinensis*), Tea (*Thea/Camellia sinensis*), Many species of Millet (*Panicum* sp.), Buckwheat (*Fagopyrum esculentum*). Several species of cabbages (*Brassica* sp.), Egg plant/Brinjal (*Solanum melongena*), Pears (*Pyrus communis*), Peach (*Prunus persica*), Common bean (*Phaseolus vulgaris*), Cow pea (*Vigna unguiculata*).

3. Hindustan centre of origin (Burma, Philippines, Java).

   Rice (*Oryza sativa*), Pigeon pea (*Cajanus cajan*), Mango (*Mangifera indica*), Coconut (*Cocos nucifera*), Sugar-cane (*Saccharum officinarum*), Banana

(*Musa* sp.), yams (*Dioscorea* sp.), Chickpea (*Cicer arientinum*), some Cotton species (*Gossypium* sp.).

4. Central Asia centre or origin (N.W. India, Afghanistan, Eastern part of former Soviet Union). It is also called the Afghanistan centre of origin.

   Garden pea (*Pisum sativum*), grapes (*Vitis vinifera*), Apple (*Pyrus malus*), Carrot (*Dauca carota*), Garlic (*Alium sativum*), Spinach (*Spinaceae oleraceae*).

5. Asia Minor centre of origin (Iran and Turkimenia).

   It is also called Near East or Persian centre of origin. Oats (*Avena sativa*), Fig (*Ficus carica*), Alfafa (*Medicago sativa*), some Wheat varieties, (*Triticum* sp.).

6. Mediterranean center of origin.

   Centre of origin for many wheats *e.g. T. durum, T. dicoccum*, Olive (*Olea europae*), Lettuce (*Lactuca sativa*), Date palm (*Phoenix dactylifera)*, Many vegetables.

7. Central American centre of origin (South Mexico and Central America). It is also called the Mexican centre of origin. Maize (*Zea mays*), Sweet potatoes (*Ipomea batata*), Paw paw (*Carica papaya*), Avacado (*Persea americana*), Sunflower (*Helianthus annus*), Guava (*Psidium guajava*), Melons (*Curcubita* sp.). Some species of cotton (*Gossypium hirsutum* and *Gossypium purpureascens*).

8. South American centre of origin (Mountainous regions of Peru, Bolivia, Equador, Colombia, Chile and Brazil). Potatoes (*Solanum tuberosum*), Tobacco (*Nicotiana tabac*um), Pineapples (*Ananas comosa*), Tomatoes (*Lycopersicon esculentum*), Cassava (*Manihot esculenta*), Rubber (*Hevea* sp.), Peanuts (*Arachis hypogaea*).

In 1935 Vavilov divided the Hindustan Centre of origin into: Java centre and Indo-Burma.

South American center of origin was divided into 3 centers: Peru, Chile and Brazilian center of origin. He also introduced the USA centre of origin.

Since some crops originated from several regions *e.g.* Cotton originated from India, Central America and Central Asia; wheat species originated from Abyssinia, Hindustan, Central Asia and Mediterranean, Some scientists prefer using the name center of genetic diversity instead of center of origin.

3.5 Microcentre

It is a small area within a center of origin that has more diversity than the whole centre. *e.g.* Coffee – Ethiopian highlands, Miraa – Kangeta in Eastern Province of Kenya.

Microcentres are important for plant introduction and evolution studies.

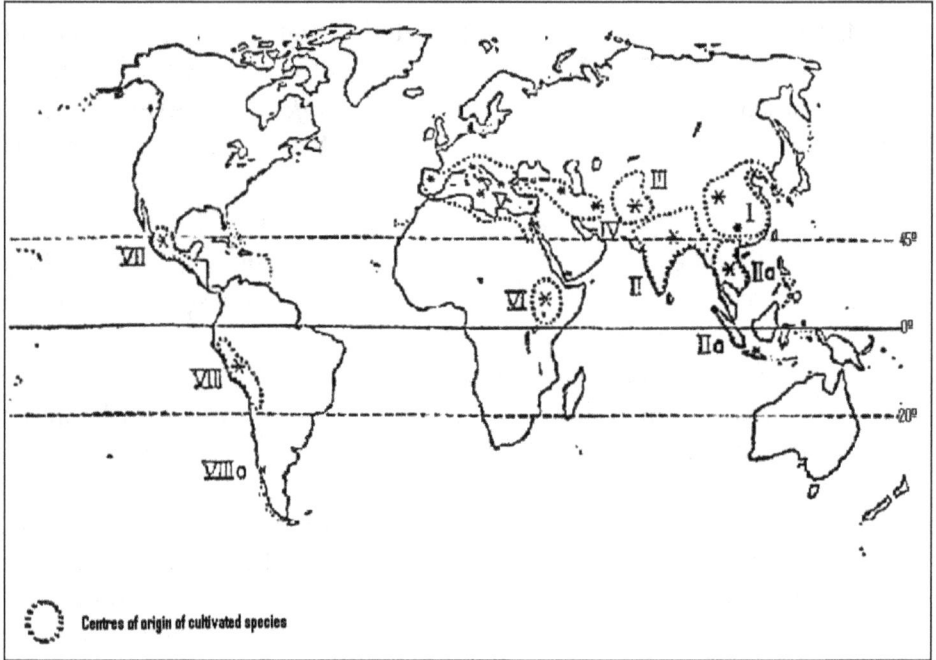

Figure 3.1: Geographical Location of World Centres of Origin
(*Source*: V.L. Chopra, 1989).

Note: Geographically, all 8 centres of origin lie between 20° South and 45° North of the Equator.

Summary

In this chapter we have discussed various centers of genetic diversity. We have also looked at the significance of centers of origin. The work of N.I. Vavilov has been illustrated. The law of homologous series in variation has been discussed. We have also identified a few crops with their origin. Microcenters and geographical location of the 8 centres have been discussed.

Self Test

1. Define a microcenter and give an example?

2. Discuss the law of homologous series in variation?

Chapter 4

Crop Genetic Resources

4.0 Introduction

Genetic resources provide genetic variation necessary for improvement of a species. It helps in crop improvement and survival of the species.

4.1 Types of Genetic Resources

Genetic resources can be obtained in several ways:

1. Wild relatives: These are plants which have continued to grow under natural conditions without man's interference. They possess important gene pools especially for adaptation to stress and resistance to pests and diseases. They are important in improving cultivated species.

2. Primary and secondary centres of origin.

3. Weedy forms: Weeds are byproducts of domestication of wild forms. They form a gap between wild populations and cultivated crops. Weeds have high fitness and unique adaptation characteristics. They possess the following traits.

   ☆ Produce many seeds
   ☆ Short dormancy period
   ☆ Strong root system
   ☆ Several methods of propagation

☆ Resistant to pests and diseases

☆ Some traits in weedy forms can be used to improve plants.

4. A known variety: Genes from known varieties can be used to improve crop plants. Genes from locally adapted varieties are preferred to exotic ones because:

☆ Exotics are expensive to conserve and have negative linkages (undesirable traits)

☆ Exotic varieties need about 3 years to acclimatize.

5. Obsolete varieties: These are varieties that were popular at one time but got replaced with time by others. They usually possess important alleles.

6. Mutations: Induced and spontaneous mutations can greatly add to the diversity of many crops. Mutations are mainly applied in breeding for disease resistance.

7. Breeding lines with particular genes and performance: These are special crops grown for their specific traits. In most cases they are not high yielding.

8. Biotechnology: It is mostly used for getting virus free crops in vegetatively propagated species. Apical meristems are usually used in tissue culture.

9. DNA Libraries: Selected DNA strands are stored in gene libraries. Later they can be incorporated into the plant genome.

10. Germplasm collection (genebank): Several centres in the world are concerned with collection and storage of various crops species *e.g.*:

☆ IBPGR – International Board for Plant Genetic Resources, Rome.

☆ IITA – International Institute for Tropical Agriculture, Nigeria.

☆ CIAT – Centre for Tropical Agriculture, Colombia.

☆ CIMMYT – International Centre for Maize and Wheat Improvement, Mexico.

☆ IRRI – International Rice Research Institute, Philippines.

☆ CIP – International Centre for Potato, Peru.

☆ ICARDA –International Centre for Agricultural Research in Dry Areas, Lebanon.

Important Agricultural Institutions in Kenya

National Gene Bank of Kenya, Muguga

Tea Research Foundation, Kericho

Coffee Research Foundation, Ruiru.

4.2 Activities in the Genetic Resource Centre

1. Exploration and Collection

A genetic resource sample should contain the fullest possible representation of genetic variability of the targeted population. Choice of sampling strategy varies from species to species and location to location.

Factors to consider when collecting germplasm:

☆ Assess the previous and current status of conservation of the crop.

☆ Know the geographical location of the area to be covered.

☆ Formulate sampling strategy.

☆ Availability of trained personnel.

Quiz: What influences priorities in germplasm collection?

☆ Economic importance of the crop.

☆ The level of genetic erosion in a given species.

☆ Need for specific germplasm.

☆ Pattern of variation of a given species.

2. Documentation

Better no record than a wrong record. Proper systematic documentation of collected and conserved material and exchange of information stimulates germplasm exchange, evaluation and utilization. Documentation can be manual or machine - based data management.

3. Conservation

Collected samples are preserved depending on:

(i) Nature of material being preserved.

(ii) Modes of reproduction.

(iii) Number and size of plants being conserved.

(iv) Length of life cycle of a species.

(v) Evolutionary status (wild or domesticated).

Conservation can be - *in situ* (natural) or *ex situ* (artificial).

In situ conservation involves protection of species in their natural environment *e.g.* Botanical gardens, National parks, Forest reserves.

Ex situ conservation involves artificial protection of species. It can be conducted in several ways.

Long-term seed storage (seed-banks) - It is the most practical and cheapest method of conservation of a species. It is based on long-term cold storage of seeds.

Note: There are 4 types of conservations depending on the duration of storage.

(i) Base collection: Long term stored materials which are disturbed only for regeneration. They are stored in sealed containers at 4-5 per cent moisture and -18 to - 20ºC temperature.

(ii) Active collections: Usually stored for 3- 5 yrs. at 8-10 per cent moisture and 5-10ºC temperature. They are regularly used in plant breeding.

(iii) Field conservation/Field Gene Banks: It is applied to plants whose seeds or vegetative parts are difficult to store. It is common in perennial tree crops.

(iv) Mass reservoirs/composite crosses: These are a large number of different parents left to breed naturally.

4. Utilization

Genetic resources are basically utilized for developing of new varieties to meet human needs. In pastures, forests and fruits species, cultivated species can easily get their counterparts in wild conditions.

4.3 Recent Genetic Resource Activities

New techniques are important not only for conservation of genetic resources but for creating new variability. Recombinant DNA is being used for screening plant breeding materials. Polymerase Chain Reaction (PCR), Restriction Fragment Length Polymorphism (RFLP) and Random Amplified Polymorphic DNA (RAPD) technologies are common. New genes can be inserted to organisms without causing disruption of the selected existing adapted genomes.

Summary

In this chapter we have discussed crop genetic resources. We have also looked at various types of genetic resources. International centers associated with genetic resources have been discussed. Methods of conservation of these resources have been discussed.

Self Test

1. Describe cryo preservation.

2. Discuss two main factors/conditions that are crucial for seed storage.

3. Describe the role played by weedy forms in plant breeding.

4. Explain various ways of getting genes of interest.

Recommended Literature

1. B. P. Singh (1995). Plant Breeding. Kalyani Publishers, 677 pages.

2. V. L. Chopra (1989). Plant Breeding: Theory and Practice. Oxford Publishers. 471 pages.

3. A. Dafni. (2001). Field methods in pollination ecology. University of Haifa, Israel. pp. 17-30.

4. J. M. Poehlman (1959). Breeding field crops. University of Missouri, 427 pages.

5. R. W. Allard (1960). Principles of plant breeding. John Wiley and Sons, Inc. California. 485 pages.

6. D. J. Van de have (1979). Heterosis in plant breeding, proceedings of the 7th Congress of EUCARPIA. Budapest. 365 pages. Y.L. Gushov, A. Fuks and P. Valichek (1999). Breeding and seed production of cultivated crops. Moscow. ISBN 5-209-00964-5. 536 pages.

7. F. V. Guliaev and Y.L. Gushov (1987). Breeding and seed production of cultivated crops. Moscow. 447 pages.

Chapter 5

Modes and Mechanisms of Reproduction in Plants

5.0 Introduction

Pollination is transfer of pollen grains from anthers to stigma. It is divided into three types: Self pollination, cross pollination and often cross pollinated.

5.1 Pollination Biology

5.1.1 Mechanisms that Encourage Self – Pollination (Autogamy)

Self- pollination is when pollen from anthers of one plant falls on the stigma of the same flower. It also includes *geitonogamy* when pollen from a flower falls on stigmas of other flowers on the same plant, *e.g.* in maize.

1. Hermaphrodite (monoecy) - staminate and pistillate flowers are on the same plant, either in the same inflorescence *e.g.* mango, banana, coconut or in different inflorescence *e.g.* maize, grapes, cassava and date palms.

2. Cleistogamy – When flowers don't open, this eliminates foreign pollen from reaching the stigma. It is found in barley, some wheat and some grasses.

3. Chasmogamy – flowers open only after pollination has taken place *e.g.* in some wheat, barley, rice.

4. Homostyly – location of stigma and anthers at the same level in a flower.

5. Muturation of stamen and pistil almost at the same time.

6. Self- compatibility (SC)-The ability of pollen tube to germinate on the stigma of the same flower to form seeds.

Note: Autogamy leads to homozygosity. Populations in self – pollinated species are highly homozygous and don't experience inbreeding depression. Breeding in autogamous crops is done to develop homozygous varieties.

5.1.2 Mechanisms that Encourage Cross-Pollination (Allogamy)

Cross- pollination is when pollen from one flower falls on the stigma of a flower in a different plant. Transfer can be done by wind – anemophily, water-hydrophily, insects-entomophily *etc.*

1. Dioecy – Male and female flowers are found on different individual plants *e.g.* papaya, spinach, date, asparagus. A female population must have a mixture of male plants.

2. Dicliny – Flowers are either staminate or pistillate. Sometimes it can be induced by emasculation or chemical means *e.g.* in maize.

3. Dichogamy – Maturation of stamens and pistil at different times. It can be:

   Protandry – Pollen is released from anthers before stigma becomes receptive, or

   Protogyny – Stigmas become receptive before pollen is released from anthers.

4. Heterostyly – Difference in length of stamens and pistils.

   (i) Distyly - A flower has long style and short stamen or short style and long stamens.

   (ii) Tristyly – The style is either short, middle or long in relation to the length of stamens.

5. Enantiostyly – Flowers have styles inclined either to the left or right of the floral axis.

6. Self incompatibility (SI) – Failure of pollen to fertilize the stigma of the same flower or other flowers on the same plant. It is a very effective way of preventing self pollination. SI can be sporophytic or gametophytic.

7. Male sterility – Absence of functional pollen grains in an hermaphrodite plant. It is usually induced by man. It can be genetic or cytoplasmic.

8. It is necessary to note that Lucerne and Alfafa stigmas have waxy film which becomes receptive only when bees break it.

Note: Allogamous crops are highly heterozygous and experience inbreeding depression when selfed. Selection is done in later generations.

Recommended isolation distance should be observed when breeding cross-pollinated crops.

5.2 How to Determine the Mode of Pollination

1. Flowers tell a lot about the mode of pollination. Mechanisms such as dioecy, monoecy, protogyny, cleistogamy clearly indicate the modes of pollination.

2. Isolating single plants and counting the number of seeds produced under isolation. Failure to set seed may indicate that a species is cross-pollinating while setting of seeds may indicates self – pollination.

3. The level of loss of vigour due to inbreeding indicates if a plant is self or cross pollinated.

4. Insect pollinated flowers are bright and have good scent.

5. Cross – pollinating plants produce more pollen than self- pollinating plants. They usually have modifications for their transfer *e.g.* feathery and light pollen grains.

5.3 Significance of the Mode of Reproduction

Modes of reproduction and pollination are very important in plant breeding because they determine:

☆ The genetic constitution of a species in population.

☆ Method in pollination control.

☆ Stability of varieties after release.

1. Genetic Constitution of a Species

Cross-pollinated crops are highly heterozygous, show inbreeding depression when selfed, consists of one or many open pollinated varieties. Self-pollinated crops are homozygous, do not experience inbreeding depression when selfed, consist of a single genotype. Vegetatively propagated crops are homozygous and can be used as a variety if desired characteristics are found in the mother plant. Clonal selection is applicable here.

2. Pollination Control

Breeding methods depend upon controlled mating. In most cases selected plants have to be crossed. In self-pollinating species selfing occurs naturally, while in cross-pollinated flowers have to be hand-pollinated and protected from foreign pollen.

3. Stability of Varieties after Release

In self-pollinated crops there is a constant genetic constitution because they are homozygous. Farmers may plant same seeds repeatedly as long as mechanical mixture is avoided. Cross – pollinated crops must be replaced every few years because of inbreeding depression.

5.4 Sexual Reproduction in Plants

It involves alternation of generations in the life cycles. Plants have diplophase (*2n*) and haplophase (*n*). Sexual reproduction involves fusion of male and female gametes to form a zygote which develops into an embryo.

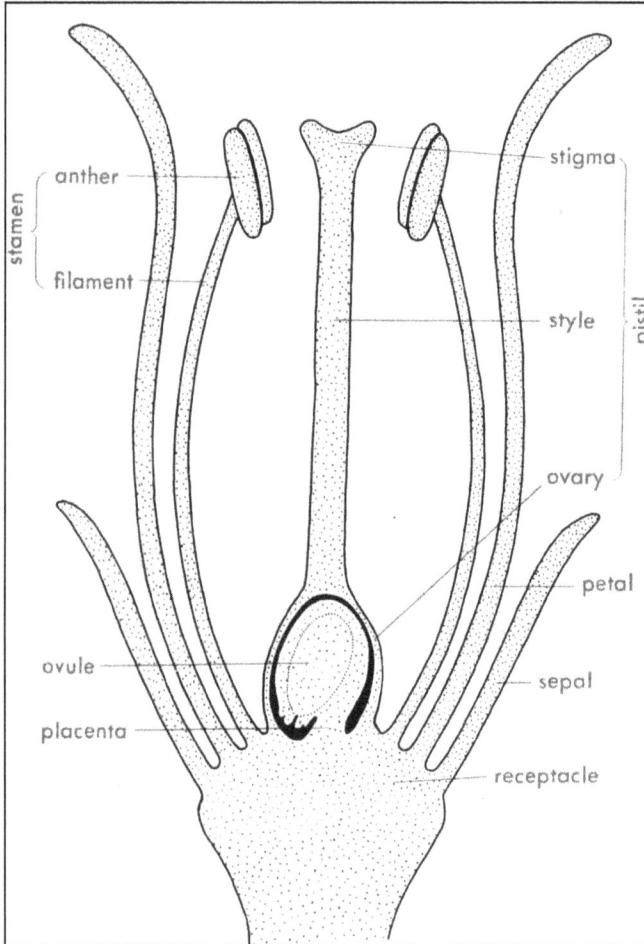

Figure 5.1: Structure of a Complete Flower
(*Source*: B.P. Singh, 1995).

A flower having both stamen and pistil is called a perfect flower (Figure 5.1). If it contains stamens but not pistils it is said to be staminate. If it contains pistils and no stamens it is said to be pistillate. Monoecius species have staminate and pistillate flowers on the same plant *e.g.* maize, coconut, castor, beans. Dioecious species have staminate and pistillate flowers on different plants *e.g.* pawpaw, date palm.

5.4.1 Formation of Male Gametes (Microgametogenesis)

Each anther has 4 pollen sacs which contain many pollen mother cells (PMC). Each PMC undergoes meiosis to produce 4 haploid cells (microspores) in a process called *microsporegenesis* (Figure 5.2).

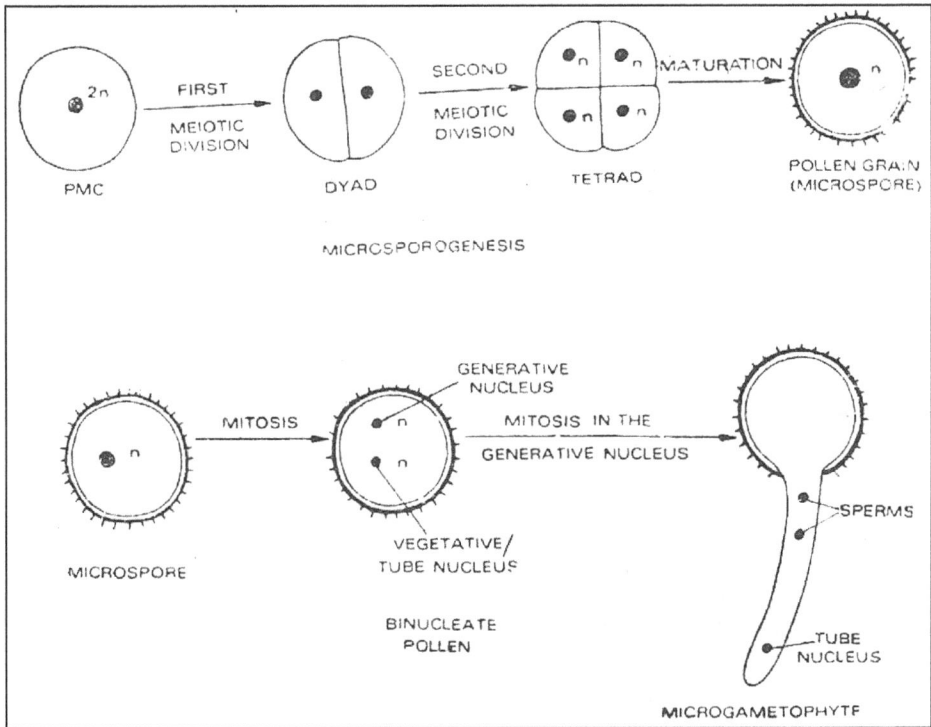

Figure 5.2: Generalised Scheme of Microsporogenesis.

Microspore nucleus divides mitotically to produce generative and vegetative nucleus. As the anther matures pollen sacs open to release pollen grains on the stigma, the vegetative nucleus grows into a tube while generative nucleus undergoes a mitotic division to form 2 male gametes or sperms.

5.4.2 Formation of Female Gametes (Megagametogenesis)

This takes place in the ovules inside the ovary. A single cell in the ovule differentiates to form megaspore mother cell. It undergoes meiosis to produce 4 haploid megaspores out of which 3 die leaving one functional megaspore per ovule (Figure 5.3).

The nucleus of a functional megaspore divides mitotically to form 4 or more nuclei. The number of nuclei and arrangement varies from species to species. Most crop plants have megaspore undergoing 3 mitotic divisions to form 8 nuclei (Figure 5.4).

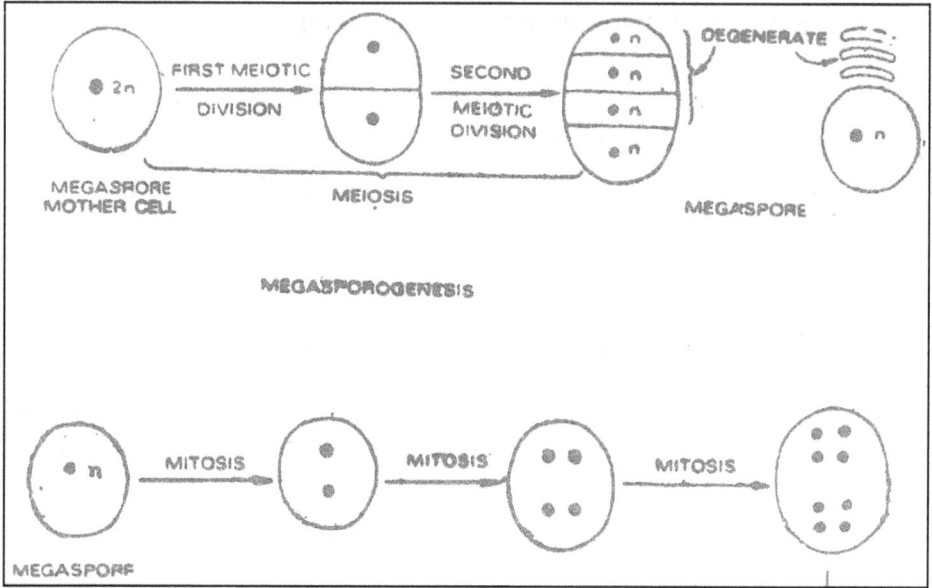

Figure 5.3: Generalised Scheme of Megagametogenesis.

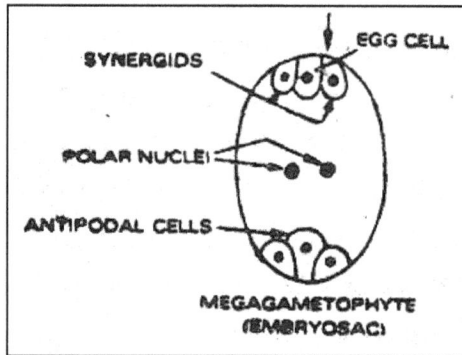

Figure 5.4: Megagametophyte (Embryosac).

Three of the nuclei move to micropylar pole to produce the central egg cell surrounded by 2 synergid cells. The other 3 nuclei move to the chalazal pole to form antipode at cells. Two nuclei remain at the center where they fuse to form polar nuclei. An embryo sac is generally made up of one egg cell (n), 2 synergids, 3 antipodal cells (all haploid) and one diploid secondary nucleus.

5.5 Pollination and Fertilization

Pollination is transfer of pollen grains from anthers to the stigma. Fertilization is fusion of one of the 2 sperms with the egg cell. In order for the stigma to be receptive to the pollen, it is branched, feathery or may secrete a sticky stigmatic fluid.

Pollen germinates on the stigma and a pollen tube grows through the style and enters the ovule through an opening called *micropyle*. The 2 sperms formed from generative nucleus move through the pollen tube into the embryo sac. Usually only one sperm joins with the; egg cell. Sometimes a second sperm may fuse with the polar nuclei to form triploid endosperm (Figure 5.5). This is called *triple fusion* common in maize.

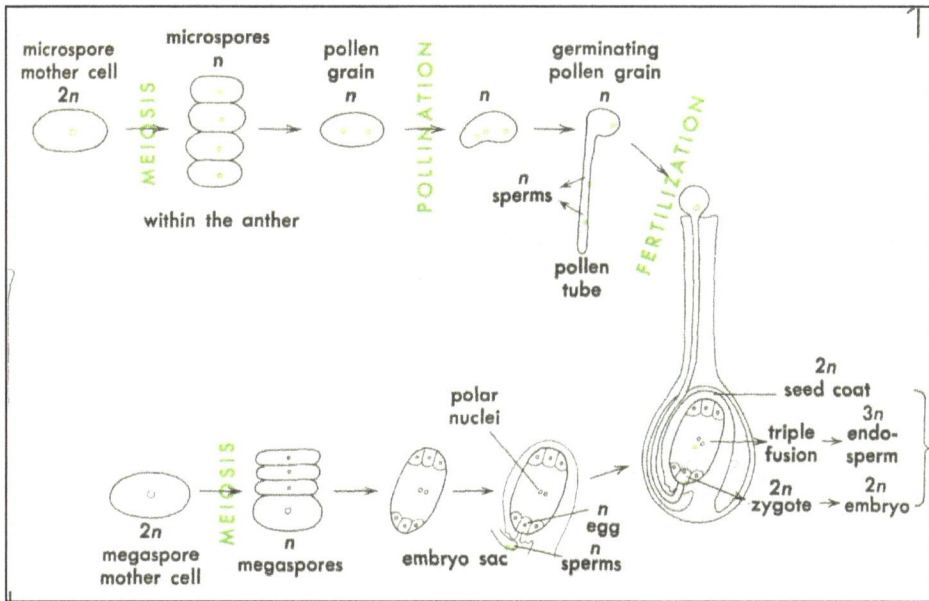

Figure 5.5: Steps that Lead to Fertilization and Eventual Formation of a Seed.

A fertilized egg develops into the embryo which on germination forms a new plant. The primary endosperm divides many times to form numerous nuclei. It is latter enclosed by cell wall to form endosperm, a tissue in which starch, oil, proteins are stored. Endosperm provides nutrition to the developing embryo. In the early stages of germination and seedling growth in cereals the larger part of seed is endosperm in legumes. It is absorbed by developing embryo so food is stored in cotyledons. The seed coat develops from integumus surrounding the ovule.

5.6 Asexual Reproduction in Plants

New plants develop from vegetative parts may arise from embryos that develop without fertilization. Vegetative reproduction is where new plants develop from a portion of plant body through modified underground and sub-aerial stems, and through bulbils.

Underground stems serve as storages organs and contain many buds.

Tuber – Potato, Bulb – Onions, Garlic, Rhizome - Ginger, Turmeric, Corm – Arrowroots, Sub – aerial stems include: runner, stolon, suckers.

Bulbils – modified flowers that develop into plants directly without formation of seeds. *e.g.* the lower flowers in the inflorescence of garlic naturally develop into bulbils.

Artificial vegetative reproduction - stem cuttings in cassava, sugar cane, grapes, roses. In many species sexual reproduction occurs naturally but for certain reasons, vegetative reproduction is more desirable.

Apomixis – formation of embryo without fertilization. Apomixis has two main types: *Parthenogenesis*- development of an embryo from egg cells without fertilization *e.g.* Nicotiana, maize.

Apogamy – Development of embryo from synergids or antipodal cells without fertilization *e.g.* in *Allium* species.

Note: Asexual reproduction produces progeny exactly identical to their parents. This preserves the genotype of an individual without change.

Clonal selection is the most effective method of breeding vegetatively propagated crops.

5.7 Incompatibility Systems in Plants

Incompatibility is dictated by a genetic system in both male and female plants. A multiple allele system denoted as S governs this system. Incompatibility can be expressed in one of the 2 systems, either gametophytic or sporophytic.

5.7.1 Gametophytic Incompatibility System

This results when pollen grain and stigma have an allele in common. Gene S has several alleles such as $S_1, S_2, S_3, S_4, \ldots \ldots$ But the number of alleles varies with species. Gametophytic incompatibility system is found in Nicotiana, Lycoperscon, Solanum, Trifolium *etc.* (Figure 5.6).

In gametophytic incompatibility systems a homozygous S allele condition is not possible through crossing.

X- ray or gamma-rays temporarily suppresses incompatibility.

5.7.2 Sporophytic Incompatibility System

It is governed by the genotype of the plants which pollen is produced and not by the genotype of the pollen. In microsporogenesis all pollen regardless of genotype retains the phenotypic response of the dominant allele in the male diploid tissue. *e.g.* an $S_1 S_2$ male would produce pollen with S_1 phenotype even though some were of S_2 genotype. There is no dominance expressed by the female genotype. There is a hierarchy of dominance in the order $S_1 > S_2 > S_3 > S_4 \ldots \ldots$ (Figure 5.7).

Unlike the gametophytic system, sporophytic allows some homozygosity of the S alleles.

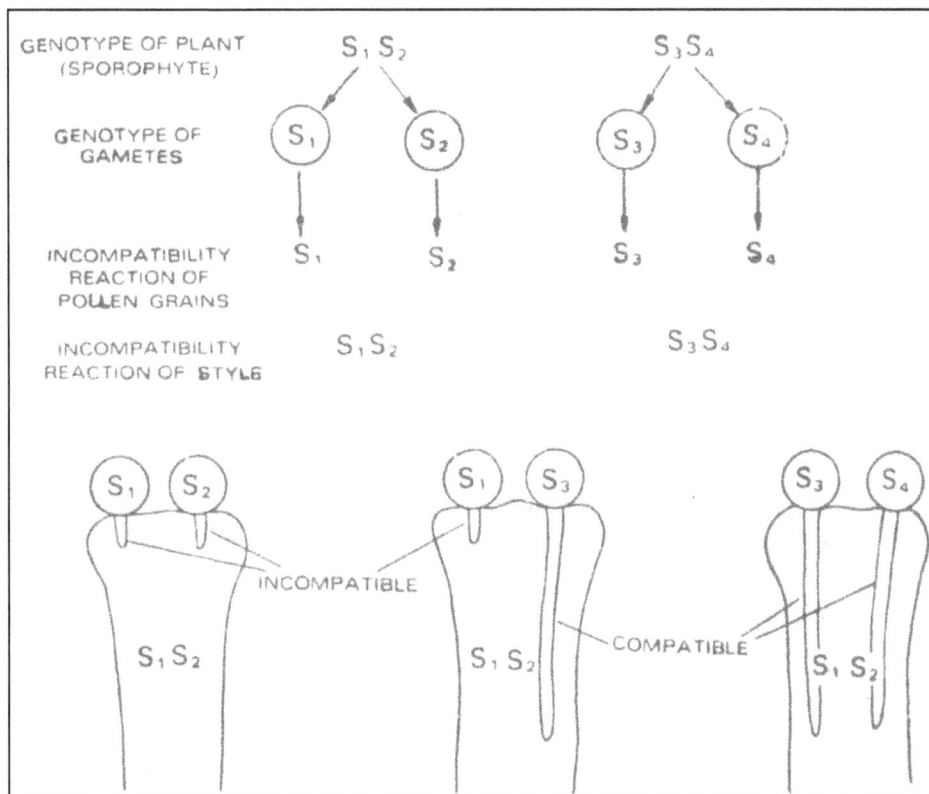

Figure 5.6: Gametophytic Incompatibility System.

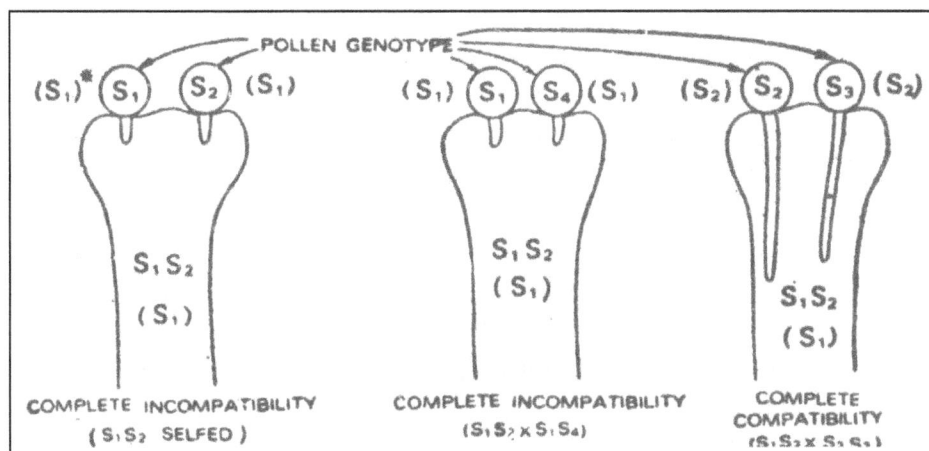

Figure 5.7: Sporophytic Incompatibility System.

Quiz: What progenies do you expect from:

a) Gametophytic incompatibility system with:

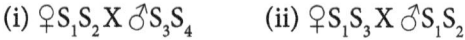

    (i) $♀S_1S_2 \times ♂S_3S_4$ (ii) $♀S_1S_3 \times ♂S_1S_2$

b) In sporophytic incompatibility system with:

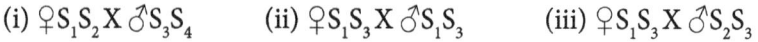

    (i) $♀S_1S_2 \times ♂S_3S_4$ (ii) $♀S_1S_3 \times ♂S_1S_3$ (iii) $♀S_1S_3 \times ♂S_2S_3$

5.8 Mechanisms of Self-incompatibility

They are quite complex and poorly understood.

(a) Pollen - stigma interaction: Pollen reaches the stigma but doesn't germinate due to poor receptive surface, too dry or wet surface and sometimes antigenic reactions.

(b) Pollen tube - style interaction: Pollen tube germinates but becomes retarded within the style due to production of inhibitory proteins and polysaccharides.

(c) Pollen tube - ovule interaction: Pollen tube reaches the ovule and effects fertilization but the ovules die at a very early stage of development.

5.9 Methods of Overcoming Self-incompatibility

1. Bud pollination – This is applying mature pollen to immature non-receptive stigma about 1-2 days before anthesis.

2. Surgical techniques – This is removal of part of the stigma or whole style to allow for pollen entrance. Sometimes the whole style may be removed and the pollen grains are directly deposited on the ovules.

3. End of season pollination- In some species incompatibility lessens towards the end of flowering season.

4. High temperature – In some species *e.g.* Trifolium, Lycopersicon, Brassica. Exposure of pistils to temperature up to 60°C induces fertility.

5. Grafting - Grafting of a branch onto another is reported to reduce self-incompatibility.

6. Double Pollination – In some species self, incompatible matings become possible when incompatible pollen is applied as a mixture with compatible.

7. Irradiation – Acute irradiation with X-ray or gamma rays induces temporary loss of self –incompatibility.

8. Treatment of pistils with phytohormones.

5.10 Male Sterility

It is when pollen grains become nonfunctional. A plant may either miss the stamens or have deformed pollen. It can be genetic, cytoplasmic or cytoplasmic–genetic.

Note: Significance – Male sterility is used for commercial production of hybrid seed in maize, cotton, sorghum, millet, cabbages without manual emasculation or detasseling.

(a) Genetic Male Sterility (GMS)

It is governed by a recessive gene ms. Male sterile alleles may arise spontaneously or may be induced. A male sterile line is maintained by crossing it with heterozygous male fertile plants (Figure 5.8).

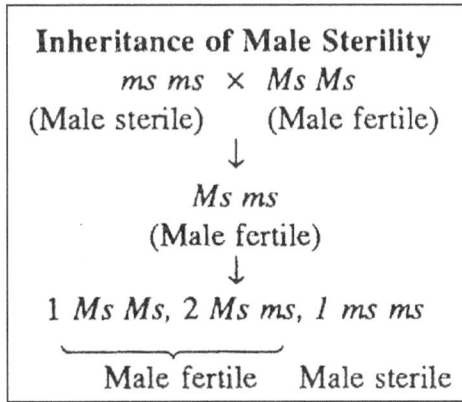

```
Inheritance of Male Sterility
        ms ms  ×  Ms Ms
   (Male sterile)      (Male fertile)
                  ↓
               Ms ms
            (Male fertile)
                  ↓
     1 Ms Ms, 2 Ms ms, 1 ms ms
     ⌣⎯⎯⎯⎯⎯⎯⎯⎯⎯⎯⌣
     Male fertile    Male sterile
```

Figure 5.8: Scheme Showing Inheritance of Genetic Male Sterility.

(b) Cytoplasmic Male Sterility (CMS)

This is male sterility determined by the cytoplasm.

It is usually inherited through the maternal line. (Figure 5.9).

In maize different types of CMS with different restorer genes have been discovered. Initials indicate the place they were first used.

CMS Type	Designation	Restorer Genes
Texas	T	Rf_1, Rf_2
Maldova	M	Rf_3
Bolivia	B	Rf, Vg_2
Paraguay	C	Rf_4, Rf_5, Rf_6

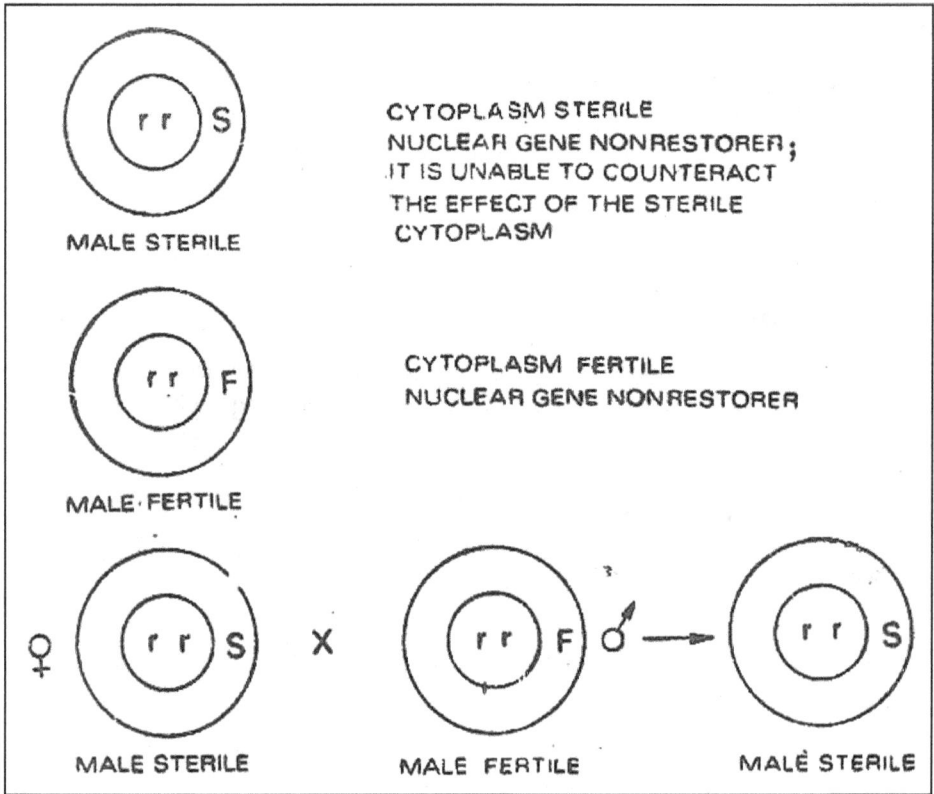

Figure 5.9: Illustration of Cytoplasmic Male Sterility.

(c) Cytoplasmic Genetic Male Sterility (CGMS)

This is sterility determined by interaction between nuclear and cytoplasmic genes. The gene for restoring fertility is well known and denoted as *R* (restorer). Plants are male sterile in presence of recessive homozygous rr. Presence of *RR* or *Rr* in the nucleus restores fertility. (Figure 5.10).

Summary

In this chapter we have discussed mechanisms that encourage autogamy and allogamy. We have also looked at methods of determining the mode of plant pollination. Microsporogenesis and megagametogenesis have been discussed. We have also learnt various incompatibility systems in plants. Male sterility in plants has also been discussed.

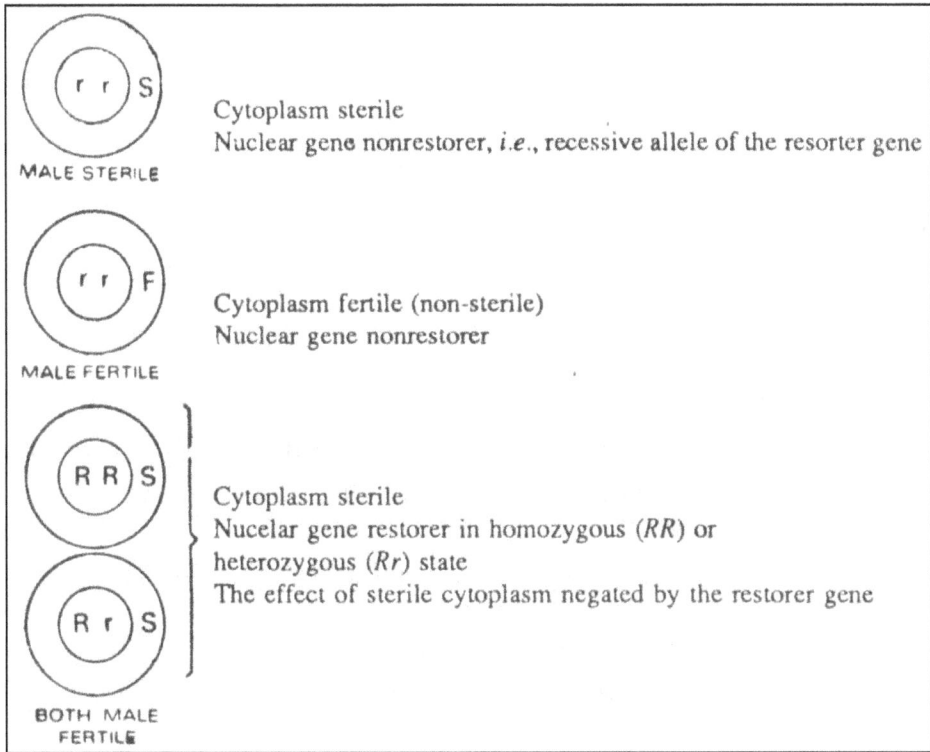

MALE STERILE

Cytoplasm sterile
Nuclear gene nonrestorer, *i.e.*, recessive allele of the resorter gene

MALE FERTILE

Cytoplasm fertile (non-sterile)
Nuclear gene nonrestorer

BOTH MALE FERTILE

Cytoplasm sterile
Nucelar gene restorer in homozygous (RR) or heterozygous (Rr) state
The effect of sterile cytoplasm negated by the restorer gene

Figure 5.10: Scheme Illustrating Cytoplasmic-Genetic Male Sterility.

Self Test

1. Explain three main factors to consider when choosing a plant breeding method.

2. What is the main significance of male sterility in seed production?

3. Give 3 main differences between sporophytic and gametophytic incompatibility systems in plants.

4. What offspring(s) would you expect between a female plant which is cytoplasmic male sterile with nuclear non-restorer genes crossing with cytoplasmic male fertile plant with heterozygous nuclear restorer genes? Show your illustrations.

5. Giving examples, define enantiostyly.

6. Describe dichogamy and its manifestation.

7. Explain the main differences between breeding self from cross pollinated crops.

Further Reading

1. B. P. Singh (1995). Plant Breeding. Kalyani Publishers, 677 pages.

2. V. L. Chopra (1989). Plant Breeding: Theory and Practice. Oxford Publishers. 471 pages.

3. A. Dafni. (2001). Field methods in pollination ecology. University of Haifa, Israel. pp. 17-30.

4. J. M. Poehlman (1959). Breeding Field crops. University of Missouri, 427 pages.

5. R. W. Allard (1960). Principles of plant breeding. John Wiley and Sons, Inc. California. 485 pages.

6. D. J. Van de have (1979). Heterosis in plant breeding, proceedings of the 7th Congress of EUCARPIA. Budapest. 365 pages.

7. Y. L. Gushov, A. Fuks and P. Valichek (1999). Breeding and seed production of cultivated crops. Moscow. ISBN 5-209-00964-5. 536 pages.

8. F. V. Guliaev and Y.L. Gushov (1987). Breeding and seed production of cultivated crops. Moscow. 447 pages.

Chapter 6

Techniques in Breeding Field Crops

6.0 Introduction

Several techniques are used in plant breeding such as emasculation, bagging, tagging and pollination.

6.1 Emasculation

Emasculation is removal of stamens or anthers or killing pollen grains of a flower without affecting the female reproductive organs. The main purpose of emasculation is to prevent self-fertilization. In dioecious plants male plants are removed while in monoecious plants with male and female flowers in different inflorescences *e.g.* maize, castor, the male inflorescence is removed. In bisexuals flowers hand emasculation must be done. Emasculation method depends on flower size, amount of seed needed, number of seed set per fruit *etc.* In species with large flowers hand emasculation is done *e.g.* cotton. In species with small flowers hand emasculation is difficult, tiring and time consuming. When large amounts of hybrid seed are needed then hand emasculation is impractical. The efficiency of emasculation is tested by bagging the emasculated flowers and checking if pollination takes place. Emasculation can be done in several ways:

(a) Hand Emasculation

It is usually done in flowers with relatively large floral parts before anthers mature and stigma becomes receptive.

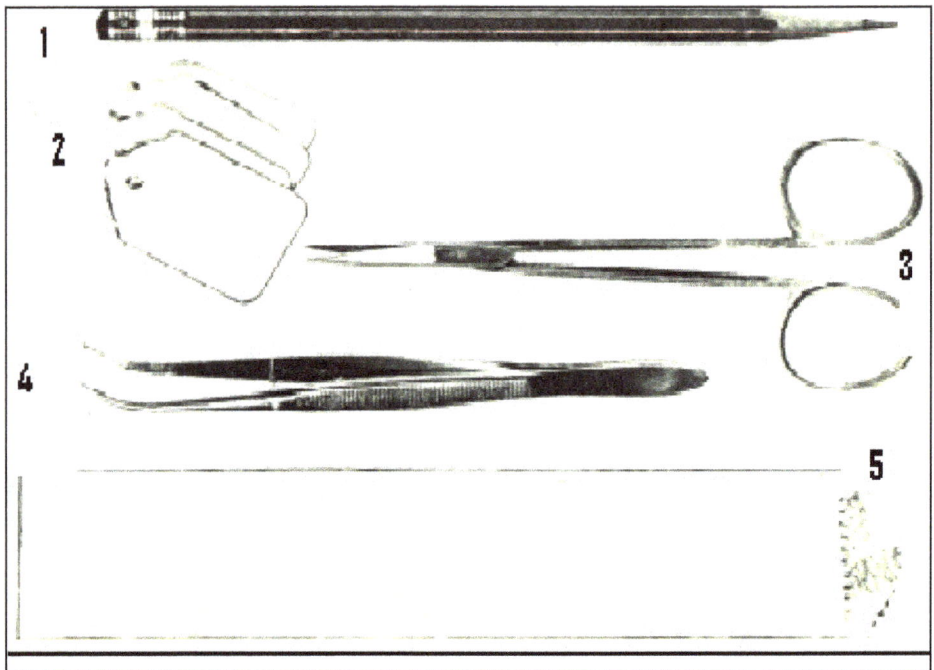

Figure 6.1: Some Tools Used in Breeding of Field Crops.
(1: Pencil; 2: Tags; 3: Pair of scissors; 4: Pair of forceps; 5: Bag).
(*Source*: J. M. Poehlman, 1959).

Most stigmas are highly receptive during morning hours when the flower opens. Evening emasculation one day before flowers mature is known to give best results. Care must be taken to remove all the anthers without breaking the female parts.

Procedure: Corolla of selected flower is fully or partly opened to expose the anthers as shown in Figure 6.2.

In cereals 1/3 of empty glumes are clipped off with scissors. In wheat the highest and lowest florets are clipped off leaving about 4 florets per spikelet. Scissors are used to chop the top part and forceps are used to remove the anthers. This is followed by bagging follows.

(b) Suction Method

It's not a very reliable method because it allows some degree of selfing. It is mostly useful in species with small flowers. It is usually done in the morning by removing petals with forceps then sucking the pollen from anthers. Washing the stigma after sucking helps in lessening self pollination.

(c) Hot Water Emasculation

Pollen grains are more sensitive to stimuli than the pistils. This advantage is taken to kill them leaving the female parts undamaged.

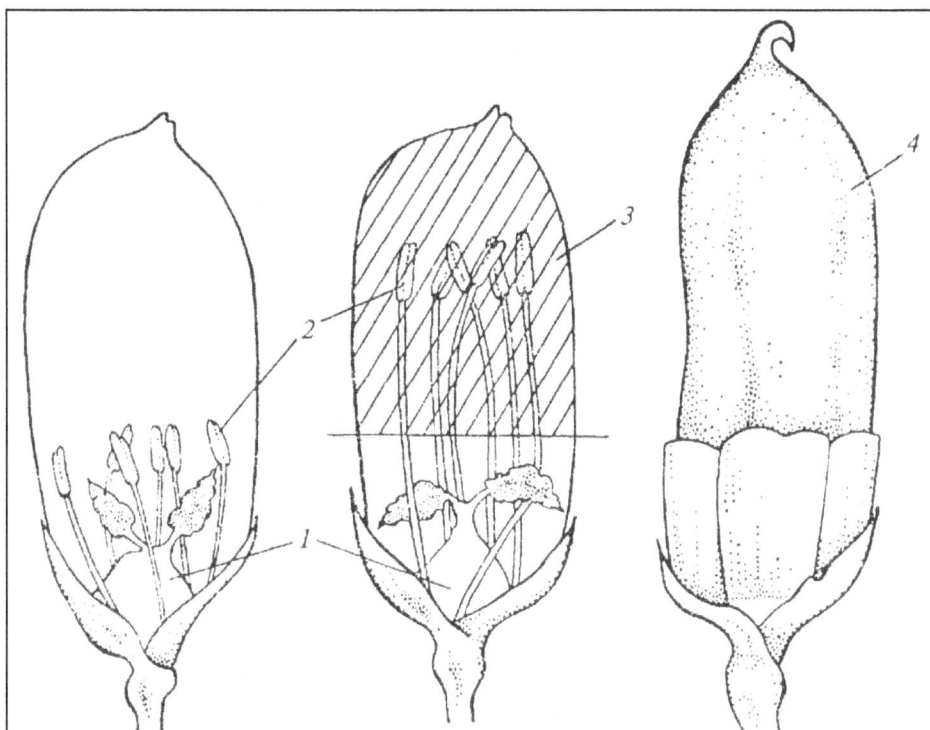

Figure 6.2: Steps in Flower Emasculation.
(1: Ovary; 2: Anthers; 3: Film formed after emasculation).
(*Source*: Y.L. Gushov, 1999).

The temperature and duration of water treatment varies from one crop to the other *e.g.* sorghum requires 42-48°C for 10 minutes. Rice requires 40–44°C for 10 minutes.

Hot water is generally carried to the field in flasks and a whole spike is inserted. It is a very effective method of killing all pollen grains.

(d) Alcohol Treatment

Flowers are inserted into alcohol of a given concentration for a given period then rinsed in water. Any prolonged time leads to killing of female parts. It's not a common method of emasculation. In clover immersion of flower in 57 per cent alcohol for 10 minutes is effective.

(e) Cold Treatment

It is not as effective as hot water treatment. Cold water kills pollen without damaging the pistils. Keeping wheat plants at 0–2°C for 15–24 hours kills the pollen grains. Rice requires 0–6°C.

(f) Use of Male sterility

It is a very effective way of ensuring nonfunctional pollen in a flowers.

6.2 Bagging

This is enclosing flowers or inflorescences in suitable bags to prevent random cross-pollination. Bags can be paper or cloth. They are usually tied at the base of inflorescence with a thread. Sometimes bagging is associated with poor seed set because of higher temperatures and humidity that encourage fungal growth on the fruit or spike.

6.3 Tagging (Identification)

Emasculated flowers are tagged just after bagging to avoid confusion. Tags are of different sizes and must contain the following information: Date of emasculation, date of pollination, name of female and male parents, breeder's name or initials. The female parent appears before the male parent. Tags must remain on the plant till harvesting and even after.

6.4 Pollination

This involves collecting pollen from dehisced anthers and dusting this pollen onto the stigmas of emasculated flowers. It may be done in several ways:

Figure 6.3: Maize Tassel Placed Next to an Ear during Pollination.

☆ Collecting pollen in a bag and dusting on female inflorescence.

☆ Collecting anthers and applying pollen with use of a brush, forceps, tooth pick *etc.*

☆ Placing an anther over the stigma where it bursts.

☆ Getting a whole male inflorescence and placing it next to a female one *e.g.* in maize as shown in Figure 6.3.

☆ Insects are used in red clover, alfalfa cabbage. They are enclosed in a cage containing the cabbage pollinator crops. It is commonly used in cabbages.

The procedure depends on species, structure of the flower and mode of pollination. In autogamous populations plants are left to follow the natural mode of pollination and seeds are harvested.

Summary

In this chapter we have looked at various techniques used in breeding field crops. We have also discussed methods of emasculation. Bagging, tagging and pollination have been discussed.

Self Test

1. Describe how to ensure the accuracy of emasculated flower.

2. What are the main factors to consider when emasculating a flower?

Recommended Literature

1. B. P. Singh (1995). Plant Breeding. Kalyani Publishers, 677 pages.

2. V. L. Chopra (1989). Plant Breeding: Theory and Practice. Oxford Publishers. 471 pages.

3. A. Dafni. (2001). Field methods in pollination ecology. University of Haifa, Israel. pp. 17-30.

4. J. M. Poehlman (1959). Breeding field crops. University of Missouri, 427 pages.

5. R. W. Allard (1960). Principles of plant breeding. John Wiley and Sons, Inc. California. 485 pages.

6. D. J. Van de have (1979). Heterosis in plant breeding, Proceedings of the 7th Congress of EUCARPIA. Budapest. 365 pages.

Chapter 7

Genetics of Disease, Pest and Drought Resistance in Crops

7.0 Introduction

Disease is an abnormal condition caused by an organism. An organism that causes disease(s) is referred to as a pathogen. A pathogen can be of plant, animal or microbial nature. The degree of damage caused by pathogens varies as follows:

Fungi > bacteria > viruses > nematodes > insects >others

Breeding efforts put against fungi are more than against other pathogens combined. Diseases can cause the following loses: Killing plants, reduce the total biomass, poor growth, deform leaf surface, lower the quality of product.

7.1 Disease Development

Development of any disease depends on a close interaction between the host, pathogen and environment. Interaction between a pathogen and its host is controlled by their genetic make up and goes on per gene for gene relationship.

Note: Gene for gene relationship states that: "*For every resistant gene present in a host, the pathogen has a gene for virulence. A susceptible reaction results only when the pathogen is able to match all the resistant genes present in the host with appropriate virulence genes. If one or more genes are not matched then there is a resistant reaction*".

Gene for gene relationship resembles a key and lock as shown below.

Figure 7.1: Illustration of Interaction between Host's Resistance Genes and Pathogen's Virulence Genes. (*Source*: J.M. Poehlman, 1959).

Almost every disease requires a specific range of environment for its rapid development beyond which it might not develop at all. Diseases keep changing their manner of attack due to changes in crop varieties and agricultural practices. New diseases may be introduced along with planting materials from other regions and countries. Quarantine measures are directed against such migrations of diseases and insect pests.

Note: Development of fungal diseases occurs in 4 main stages:

1. *Contact* - Landing of a pathogen on a host.
2. *Infection* - This is entrance of a pathogen in a host. It can be through natural openings like the stomata, damaged tissues and sometimes through the epidermis by use of degradative enzymes.
3. *Establishment* - The pathogen multiplies and spreads inside the host but the symptoms are not seen (incubation).
4. *Development* - High spore multiplication, symptoms appear, damage is caused, release of spores.

7.2 Disease Escape

It is when a susceptible host plant is freed from diseases due to environmental factors. In disease development both contact and infection stages are the mostly affected by environmental conditions. Poor environmental factors would adversely reduce or prevent germination of spores and infection. Disease escape can be enhanced in several ways:

☆ Planting under crop rotation

☆ Changing planting date to avoid a pathogen

☆ Balanced application of fertilizers helps in boosting the immune system of plants

7.3 Genetics of Pathogenecity

Pathogenecity is the ability of a pathogen to attack a host. It is synonymous to virulence. Virulence of a pathogen is under genetic control. In most cases more than one gene is involved in governing virulence. In many pathogens such as smuts, rusts, it has been proved that virulence is recessive to avirulence.

Physiological races - Strains of a single pathogen species that differ in their ability to attack different varieties of the same host species. Hosts resistant genes are usually not known. Varieties of a host species used to identify physiological races of a pathogen are known as *differential hosts* or *host testers*.

Pathotypes (biotypes) - Strain of a pathogen virulent towards a specific known resistant gene of the host. The resistant gene is known.

Quiz: What is the difference between a physiological race and a pathotype?

7.4 Reaction to Diseases

Reaction of a host to a disease can be grouped as follows:

1. ***Susceptible reaction:*** Genes of the host are not able to counteract those of the pathogen. As a result there is intensive disease development.

2. ***Immune reaction:*** There is contact and infection but no symptoms are seen. This can be due to host cells around the point of infection dying.

3. ***Resistance:*** It is less disease development than in susceptible. Infection and establishment take place but growth of the pathogen is restricted. This may be due to poor nutrition from the host or release of growth inhibitors.

4. ***Tolerance:*** The host is attacked by the pathogen just like the susceptible variety but there is no direct reflection on the final product. This is mostly due to attack on non-strategic organs, or attack when the plant has undergone important physiological processes or quick recovery from a disease.

Note: Most of these classifications are relative in their definition. Incase a reaction is not well defined we use scales with varying units. Explanation of the scale varies from one country to the other. Some use 0 (no disease) to 9 (most susceptible).

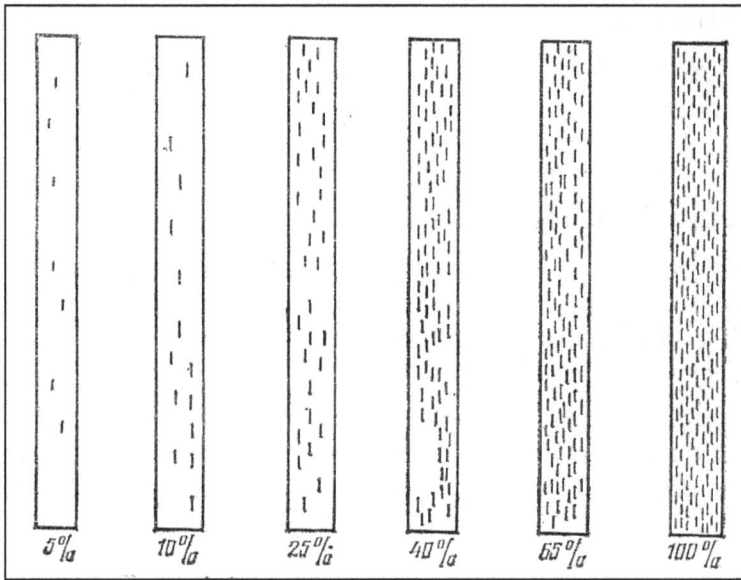

Figure 7.2: Scale for Estimating Percentage of Rust
(*Puccinia graminis tritici*) Infection on the Leaf or Stem.
(*Source*: Y. L. Gushov, 1999).

7.5 Types of Disease Resistance

(a) Vertical Resistance (VR)

It is also referred to as race- specific, pathotype specific or simply specific resistance. It is resistance to specific races of a pathogen but not to others. It is usually governed by a major (oligo) gene. A host carrying a gene for vertical resistance is attacked only by the pathogen which is virulent towards that resistant gene. To all other pathotypes the host will be resistant. If the virulent pathotype becomes frequent epidemics are common in the case of vertical resistance.

Vertical resistance has been found in all classes of pathogens (fungi, bacteria, viruses, nematodes, insects *etc.*).

(b) Horizontal Resistance (HR)

It is also referred to as race – nonspecific, pathotype-nonspecific, partial or general resistance. It is controlled by polygenes. It is effective against all races of a pathogens. It does not prevent development of symptoms but slows down the rate at which a disease spreads. It is sometimes very difficult to separate HR from VR.

Table 7.1: Various Types of Disease Resistance

Strain	Pathotype			Type of Resistance
	P_1	P_2	P_3	
Race 1	S	S	R	Verticle
Race 2	R	S	S	Verticle
Race 3	S	R	S	Verticle
Race 4	R	R	R	Horizontal

(c) Durable Resistance (DR)

It remains effective over a prolonged period. It is assessed only on retrospective (past) analysis. Sometimes is difficult to know if the plant possesses DR or not.

Aggressiveness (horizontal pathogenicity)

It is the level at which pathogenic strains having the same virulence genes differ in causing different amount of disease on a given host genotype.

7.6 Physiological and Morphological Mechanisms of Disease Resistance

Plants have over a long time developed mechanisms of contracting diseases such as mechanical, hypersensitivity, antibiosis and nutritional.

1. Mechanical - Several mechanical and anatomical features of host plants may prevent infection *e.g.* closed flowering habit in wheat and barley may prevent infection by spores of ovary infecting fungi. Thick cuticle layer prevents entrance of pathogens to the plant tissues.

2. Hypersensitivity - In most cases immunity of plants is due to hypersensitivity reaction. After infection the plant releases chemicals called *phytoalexins* that cause cells surrounding the point of infection to die.

3. Antibiosis (food poisoning) some plants release toxic substances that may reduce disease development, interfere with reproduction or may even kill the pathogen. A good example is gossypol in cotton. Cultivars lacking gossypol are more susceptible to boll worm and other pests than those without.

4. Nutritional - When a host plant does not fulfill nutritional requirements of the pathogen, there is poor growth and poor spore production of the pathogens.

7.7 Inheritance of Disease and Pest Resistance

Inheritance is passing on genetic information from one generation to the next. Scientists have recognized three different modes of inheritance: Oligogenic, polygenic and cytoplasmic.

(a) Oligogenic Inheritance/Vertical Inheritance

☆ Resistance is governed by one or few major genes and resistance is generally dominant to susceptible reaction.

☆ It is pathotype – specific, meaning that resistant gene is effective against some pathotypes and ineffective against others.

☆ Each gene has a large and identifiable individual effect on resistance. This means that the difference between resistance and susceptible plants is usually clear cut.

☆ Environment has little effect on the expression of an oligogene meaning that phenotypic expression reflects the genotype.

☆ Resistance is easily transferred.

☆ Resistance is inherited in the ration of 3:1 in the F_2.

Table 7.2: Inheritance of Resistance to Anthracrose in Common Bean (*Phaseolus vulgaris*)

Pathotype Strain	Bean Variety		Cross between Marrow and Robust	Ratio in F_2
	White Marrow	Robust		
Alpha	Resistant	Susceptible	Resistant	3:1
Beta	Susceptible	Resistant	Resistant	3:1

Ratio 3:1 in F_2 means that resistance is governed by a single gene with 2 alleles.

Note: To determine the mode of inheritance you are guided by the ratio of resistance and susceptible plants in the F_2.

Task. A plant breeder crossed two maize cultivars. One resistant and the other susceptible to leaf rust. He used 40 seeds of resistant and 40 seeds of susceptible plants. In the first generation he got 38 resistant and 2 susceptible. After crossing F_1 hybrids among themselves he got a total of 189 plants out of which 140 were resistant and 49 were susceptible. What mode of inheritance was involved in these crosses?

Table 7.3: Example of Oligogenic Inheritance

Cultivars/Cross	Generation	Total Plants	Observed Number	
			Resistant	Susceptible
Km -49	P_1	40	40	0
Lm - 30	P_2	40	0	40
	F_1	40	38	2
	F_2	189	140	49

Solution: F_2 has a total of 189 in the ratio 140:49 which is 3:1. This means that inheritance was oligogenic.

Table 7.4: Example of Oligogenic Inheritance

Genotype		Total Seeds	Observed	
			Resistant	Susceptible
A	P_1	60	60	0
B	P_2	60	0	60
A x B	F_1	60	58	2
(A x B) x (A x B)	F_2	260	196	64

What mode of inheritance is this?

Solution: In F_2 the ratio is 196:64 which is 3:1 meaning inheritance is oligogenic.

(b) Polygenic Inheritance

☆ Disease resistance is governed by many genes with small additive effects.

☆ No clear cut difference between resistant and susceptible plants.

☆ There is a continuous variation in phenotypic expression.

☆ Transfer of resistance is relatively more difficult than in oligogenic resistance

☆ There is a large environmental effect on gene expression. Phenotype is a result of joint action between genotype and environment.

Note: Only the genotypic contribution is inherited.

Polygenic inheritance is usually in the ratios 15:1 and 63:1 in F_2.

Example

Students studied inheritance of seed colour in beans. The F_2 generations from various crosses had red and white grains in the ratios 3:1, 15:1, 63:1. This means that seed colour was governed by one, two and three genes respectively.

Table 7.5: Gene Interaction in Polygenic Inheritance among Bean Colours

Gametes	AB	Ab	aB	ab	
AB	AABB	AABb	AaBB	AaBb	Complete heterozygosity
Ab	AABb	AAbb	AaBb	Aabb	
aB	AaBB	AaBb	aaBB	aaBb	
ab	AaBb	Aabb	aaBb	AaBb	Complete homozygosity

* Dotted lines represent complete homozygotes and complete heterozygotes respectively

A thorough observation of 15 red: 1 white showed, that red colour differed in its intensity as follows: 1 dark red, 4 medium dark red, 6 medium red and 4 light red.

(C) Cytoplasmic Inheritance

Most cellular DNA is concentrated in the nucleus but it is also found in other organelles as the mitochondria and chloroplasts. Cytoplasmic inheritance is mainly through the maternal line.

Example: Maize strains having cm-T are highly susceptible to *Helminthosporium maydis* while those with normal cytoplasm are resistant to the disease. Chloroplasts aren't wholly dependant on nuclear genes. A plant with abnormal chloroplast will have offsprings with abnormal chloroplasts.

7.8 Morphological and Physiological Factors of Insect Resistance in Crops

A plant may have one or more mechanisms to counteract insect attack. Insect resistance mainly involves morphological and physiological features of the host plants.

(A) Morphological Factors

1. Plant colour may determine the choice of an insect to colonize or settle on it *e.g.* Butterflies prefer green cabbage to red, Bollworms prefer green cotton to red.
2. Hairiness of leaves and stems is associated with resistance to many pests.
3. Presence of awns in cereals keeps off pests and birds.
4. Thick and tough plant tissues make it difficult for insects to attack a plant.
5. Narrow leathery shapes of leaves and flowers confer insect resistance than in broad succulent ones.

(B) Physiological Factors

1. Presence of nectar in a plant attracts both insects and birds.
2. Leaves of some potatoes species secrete gummy substances which trap aphids and colorado beetles.
3. High silica content in rice shoots confers resistance to shoot borer by digesting the pest's mandibular mouth parts.
4. Many tomato species have ethanol soluble compounds which are highly toxic to tomato fruit worm.

(C) Other Mechanisms of Insect Resistance

1. Antibiosis (food poisoning) *e.g.* pyrethrin confers plant resistance to many pests

2. Tolerance - A plant is attacked by insects just like the susceptible one but the product of interest is less or not affected. This is mainly due to ability of a plant to recover from damages or attack on non-strategic organs *e.g.* tassels.

3. Avoidance - Synonymous to disease escape in pathogens. This is when host plants are less susceptible to attack at the peak of insect population. This is enhanced by timed planting dates.

7.9 Mechanisms of Drought Resistance in Crops

Drought is a meteorological event which implies the absence of rainfall for a period of time, long enough to cause moisture-depletion in soil and water deficit with a decrease of water potential in plant tissues. From the agricultural point of view, drought is inadequacy of water availability, including precipitation and soil-moisture storage capacity, in quantity and distribution during the life cycle of a crop plant, which restricts the expression of full genetic potential of the plant. In agriculture, drought resistance refers to the ability of a crop plant to produce its economic product with minimum loss in a water-deficit environment relative to the water-constraint-free management. An understanding of genetic basis of drought resistance in crop plants is a pre-requisite for a geneticist to evolve superior genotype through either conventional breeding methodology or biotechnological approach.

In genetic sense, the mechanisms of drought resistance can be grouped into three categories: Drought escape, Drought avoidance and Drought tolerance.

However, crop plants use more than one mechanism at a time to resist drought. Drought escape is defined as the ability of a plant to complete its life cycle before serious soil and plant water deficits develop. This mechanism involves rapid phenological development (early flowering and early maturity), developmental plasticity (variation in duration of growth period depending on the extent of water-deficit) and remobilization of pre-anthesis assimilates to grain. Drought avoidance is the ability of plants to maintain relatively high tissue water potential despite a shortage of soil-moisture, whereas drought tolerance is the ability to withstand water-deficit with low tissue water potential. Mechanisms for improving water uptake, storing in plant cell and reducing water loss confer drought avoidance. The responses of plants to tissue water-deficit determine their level of drought tolerance. Drought avoidance is performed by maintenance of turgor through increased rooting depth, efficient root system and increased hydraulic conductance and by reduction of water loss through reduced epidermal (stomatal and lenticular) conductance, reduced absorption of radiation by leaf rolling or folding and reduced evaporation surface (leaf area). Plants under drought condition survive by doing a balancing act between maintenance of turgor and reduction of water loss. The mechanisms of drought tolerance are maintenance of turgor through osmotic adjustment (a process which induces solute accumulation in cell), increase in elasticity in cell and decrease in cell size and desiccation tolerance by protoplasmic resistance.

Unfortunately, most of these adaptations to drought have disadvantages. A genotype of short duration usually yields less compared to that of normal duration. The mechanisms that confer drought resistance by reducing water loss (such as stomatal closure and reduced leaf area) usually result in reduced assimilation of carbon dioxide. Osmotic adjustment increases drought resistance by maintaining plant turgor, but the increased solute concentration responsible for osmotic adjustment may have detrimental effect in addition to energy requirement for osmotic adjustment. Consequently, crop adaptation must reflect a balance among escape, avoidance and tolerance while maintaining adequate productivity.

Summary

In this chapter we have defined a disease. We have also discussed various stages of disease development. Mode of inheritance have also been discussed. Physiological and morphological mechanisms of disease and pest resistance have been discussed. Mechanisms of drought resistance have also been discussed.

Self Test

1. A plant breeder made 3 different crosses and found the F_2 ratios between black and white seed colour as follows respectively: (a) 286:19 (b) 190:3. Comment.

2. Describe 'gene for gene' relationship.

3. Differentiate a physiological race from a pathotype.

4. What is the difference between tolerance and susceptible reactions to disease attack?

5. Explain a reciprocal cross and its significance in plant breeding.

Recommended Literature

1. B. P. Singh (1995). Plant Breeding. Kalyani Publishers, 677 pages.

2. V. L. Chopra (1989). Plant Breeding: Theory and Practice. Oxford Publishers. 471 pages.

3. A. Dafni. (2001). Field methods in pollination ecology. University of Haifa, Israel. P. 17-30.

4. J. M. Poehlman (1959). Breeding field crops. University of Missouri, 427 pages.

5. R. W. Allard (1960). Principles of plant breeding. John Wiley and Sons, Inc. California. 485 pages.

Chapter 8

Types of Crosses

8.1 Simple Cross

Two parents are crossed to produce the F_1 as shown in Figure 8.1.

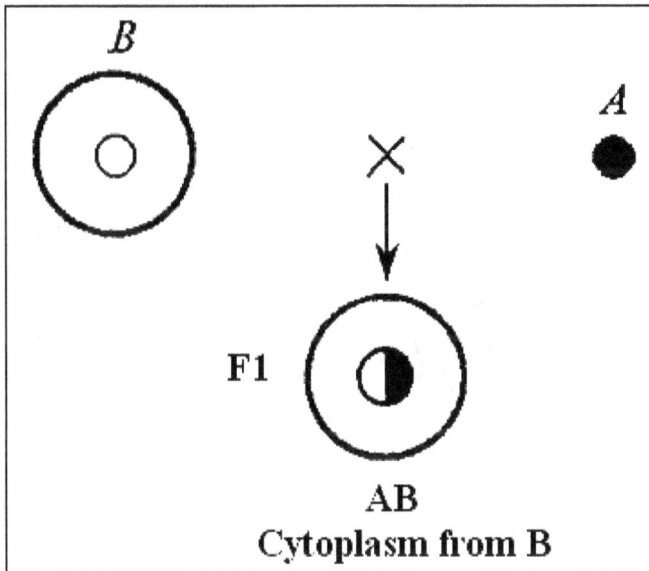

Figure 8.1: Simple Cross.

8.2 Double Cross

This is when the F_1 of two simple crosses is merged to form the F_2 (Figure 8.2).

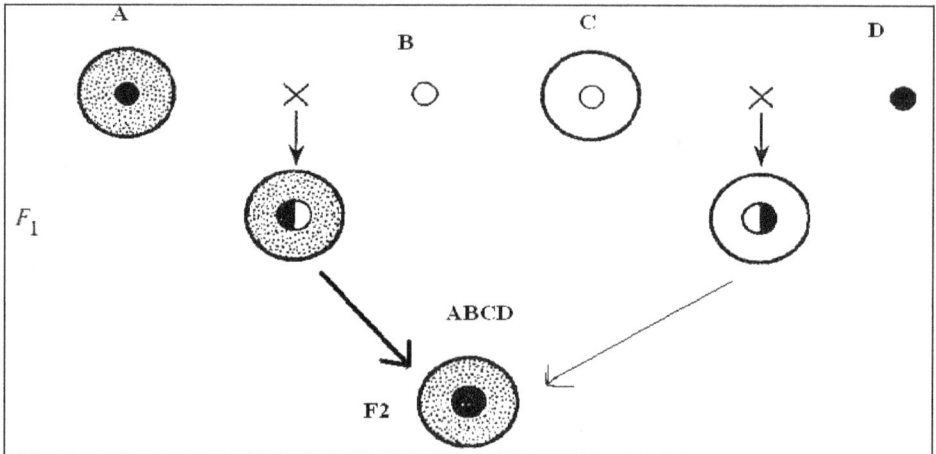

Figure 8.2: Double Cross.

8.3 Topcross

Phenotypically selected plants are intercrossed with a common pollen parent (tester) on the basis of their performance desired progenies are selected (Figure 8.3).

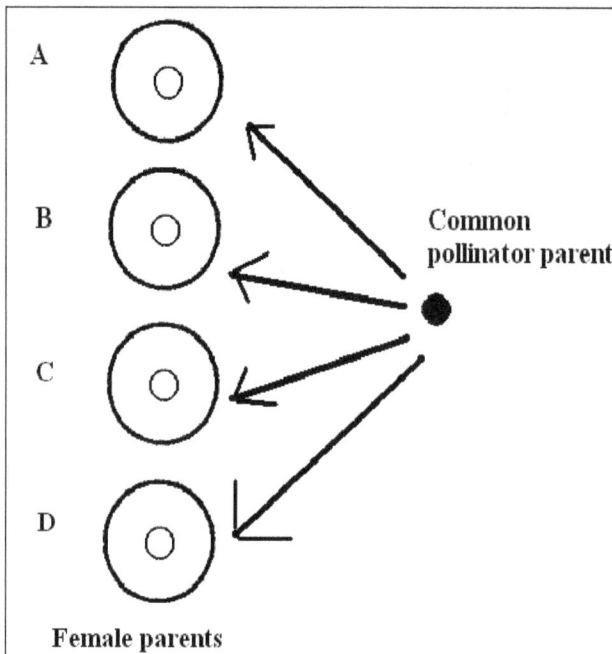

Figure 8.3: Topcross.

8.4 Reciprocal Cross

It is when each of the two parents serves a mother in one cross and a father in the next cross. It is used to determine interaction between nuclear and cytoplasmic genes (Figure 8.4).

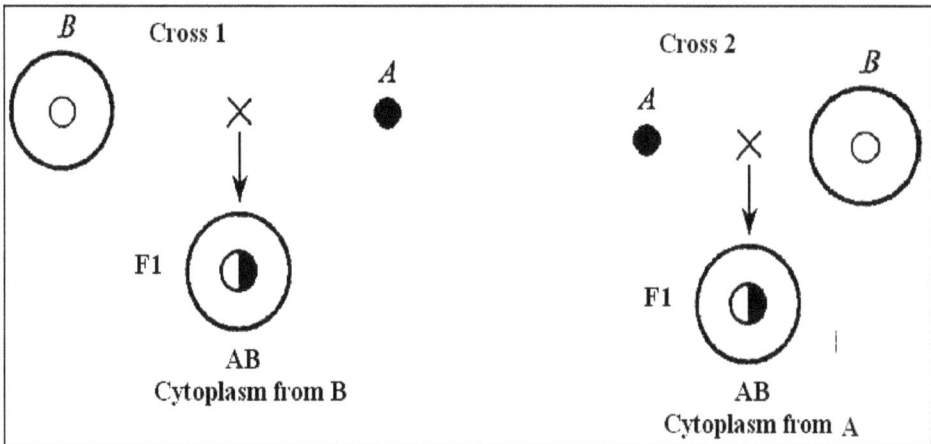

Figure 8.4: Reciprocal Cross.

8.5 Polycross

Different variants of the same mother plant are pollinated by different male parents. It is used in cross pollinated crops especially grasses (Figure 8.5).

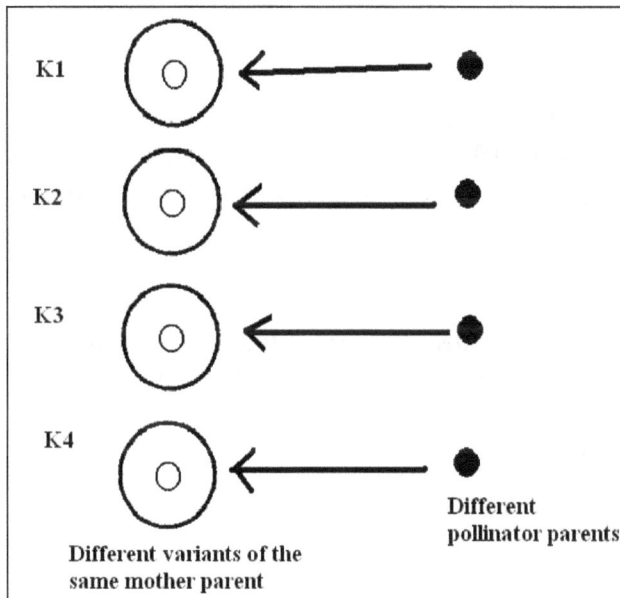

Figure 8.5: Polycross.

8.6 Diallel Cross

Every line is crossed with all possible combinations. It is used for determining specific combining ability for heterosis breeding (Table 8.1).

Table 8.1: Diallel Cross

	A	B	C	D
A	-	AB	AC	AD
B	BA	-	BC	BD
C	CA	CB	-	CD
D	DA	DB	DC	-

8.7 Backcross

It is a cross in which the progeny is repeatedly crossed with one of the parents. It is useful when some qualities must be maintained as others are added. It is mainly used in breeding for pest and disease resistance (Figure 8.6).

8.8 Conventional Cross

It involves parallel backcrossing and final merging of the hybrids. It confers several characteristics to good yielding variety within a shorter time (Figure 8.7).

Example: A high yielding wheat variety has become susceptible to wheat rust and aphids. A conventional cross can be used as follows:

Note: Gene engineering and methods of biotechnology are the most widely used nowadays, they shorten the time taken but are associated with some negative effects.

Summary

In this chapter we have identified various types of crosses used in plant breeding. We have also discussed their application in plant breeding. Breeding schemes have been illustrated.

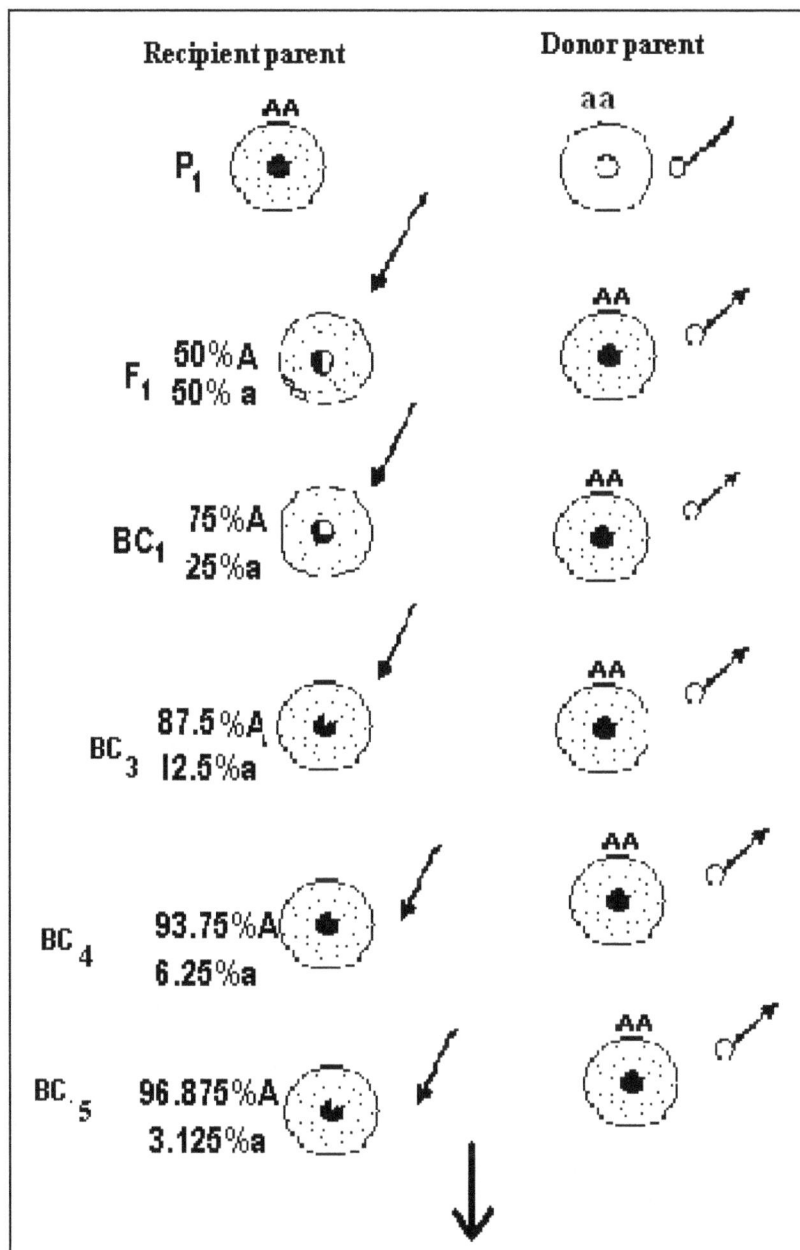

Figure 8.6: Backcross.

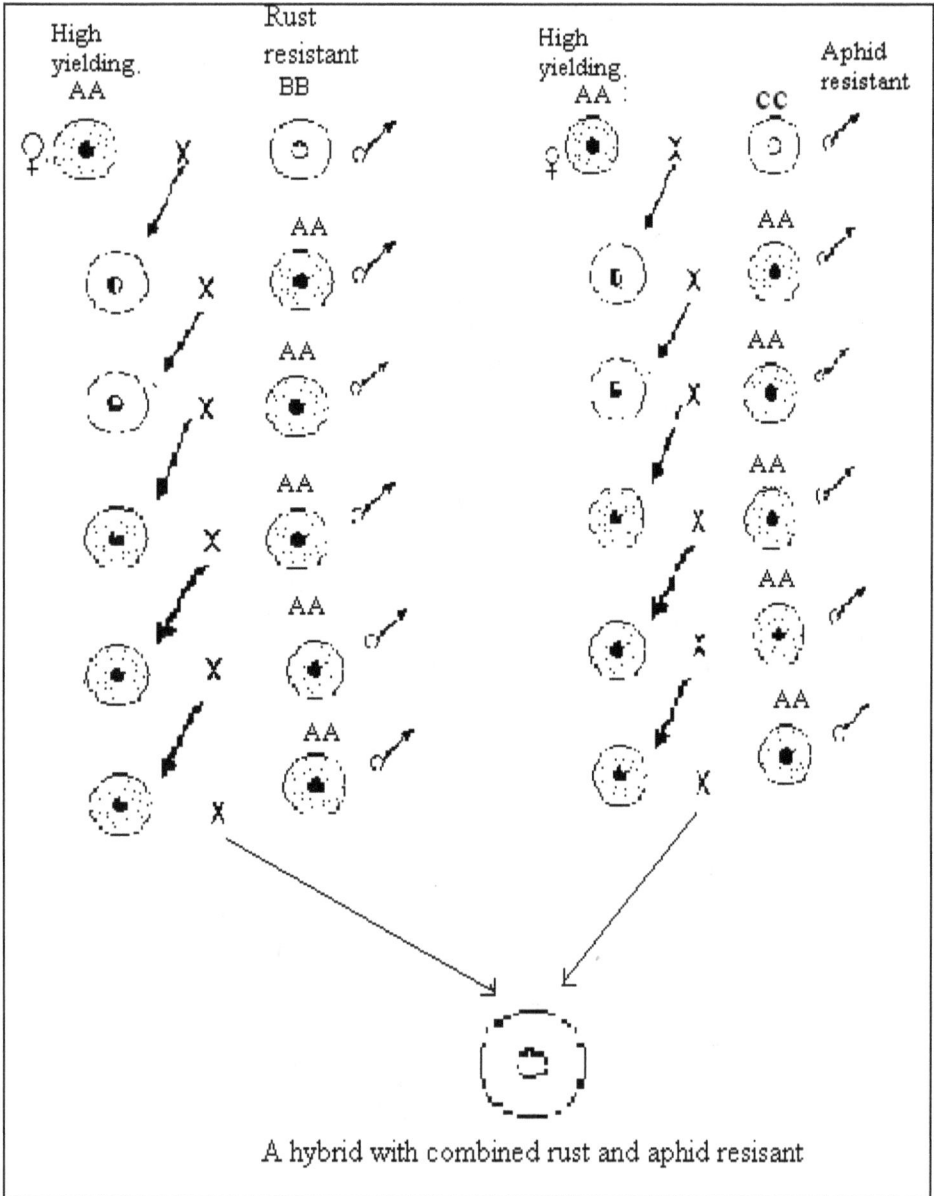

Figure 8.7: Scheme Illustrating a Conventional Cross.

Further Reading

1. V. L. Chopra (1989). Plant Breeding: Theory and Practice. Oxford Publishers. 471 pages.

2. A. Dafni. (2001). Field methods in pollination ecology. University of Haifa, Israel. P. 17-30.

3. J. M. Poehlman (1959). Breeding Field crops. University of Missouri, 427 pages.

4. R. W. Allard (1960). Principles of plant breeding. John Wiley and Sons, Inc. California. 485 pages.

5. Y. L. Gushov, A. Fuks and P. Valichek (1999). Breeding and seed production of cultivated crops. Moscow. ISBN 5-209-00964-5. 536 pages.

Chapter 9

Plant Breeding Methods

9.1 Factors to Consider while Choosing a Plant Breeding Method

In choosing the appropriate method of plant breeding, the following factors should be considered:

1. Mode of plant reproduction - Plants can have sexual, asexual or a combination of both modes of reproduction. Breeding method will entirely depend on reproduction mode.

2. Level and type of self-incompatibility and its causes.

3. Possibility of inbreeding depression when selfed.

4. Mode of inheritance of the desired trait. Oligogenic inheritance is easier than polygenic.

5. Whether the desired trait is dominant over the undesired. Breeding for a dominant trait is easier than recessive.

6. Peculiarity of the flower structure and the amount of pollen produced. Usually cross-pollinated plants produce more pollen than self pollinated.

7. Financial capability - Possibility of using green houses and off-season nurseries.

8. Reproduction coefficient - This is the number of progenies derived from a single mother plant. *e.g.* One maize seed gives about 900 seeds, one Sugarcane cutting gives about 8 shoots.

This coefficient is lowest in vegetatively propagated crops.

Note: Almost all breeding methods follow the following general steps:

☆ Creation of variation or use of the already existing variation.

☆ Selection of desired plants from a mixed population

☆ Evaluation of the performance of the new plants in different locations in comparison with the existing ones.

☆ Multiplication of planting materials and distribution to farmers.

9.2 Backcross Method

A backcross is a cross between a hybrid and one of its parents.

Objective

To improve varieties that excel in several important traits but lack an important component *e.g.* Disease resistance.

Idea

The character(s) lacking in this variety are transferred to it from a donor parent without changing its genotype but adding genes that are being transferred. The end result is a variety having the original desired characteristics and other added characteristics.

Example

KU is a high yielding wheat variety that has become susceptible to leaf rust. In order to improve and maintain this variety, it is necessary to look for resistance genes from known variety, related species, genebanks, centers of genetic diversity *etc.*

In our case, KU is the recipient (recurrent parent), while the plant carrying resistant genes is the donor (non-recurrent parent).

Recurrent parent – parent to which genes are being transferred from the donor. It is repeatedly used in the backcross.

Donor parent – parent from which genes are transferred to the recurrent parent. A donor is used only once to produce the F_1 hybrids.

Requirements of a backcross programme for resistance:

1. An agronomically suitable recurrent parent that lacks one or two genes. This parent must retain its original traits and accept new ones.

2. Suitable donor parent with enough resistant genes for transfer.

3. Character being transferred must have high heritability. Preferably controlled by one or few genes.

4. Sufficient number of backcrosses must be done to recover the original genotype of the recurrent parent. Usually 5 – 6 backcrosses are enough for this purpose.

5. Presence of artificial epidemics to identify and screen the resistant plants. Epidemics can be pathogens or insects.

Consequences of Repeated Backcrossing

After every backcross with the recurrent parent the proportion of donor genes reduces by half while those of recurrent parent increases by half (Figure 9.1).

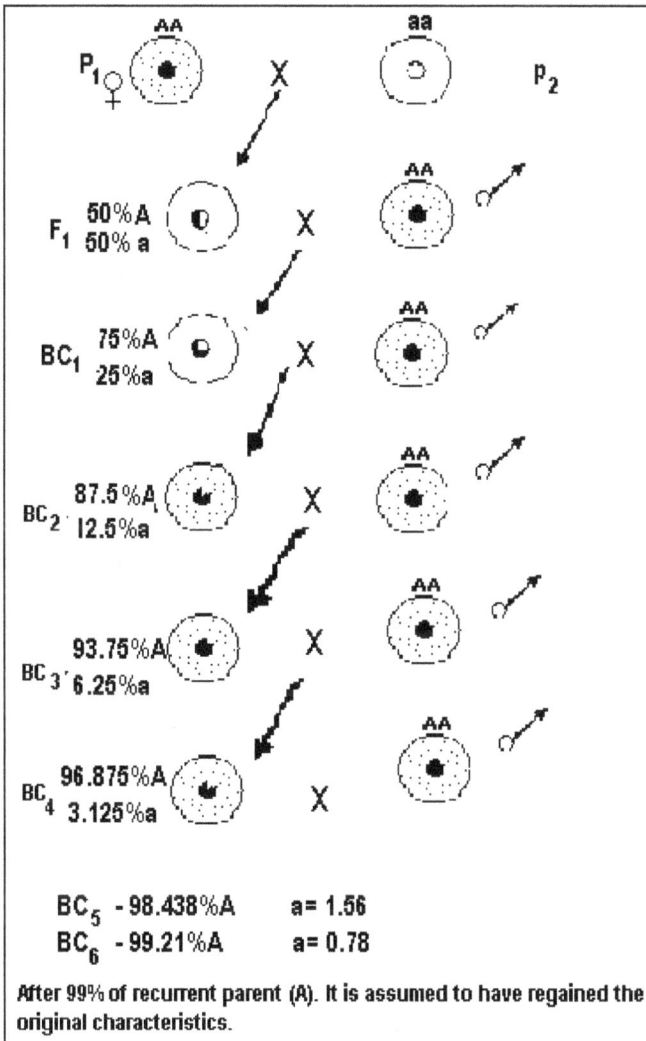

Figure 9.1: Genetic Consequences of Backcrossing.

Note: To maintain the character under transfer selection must be done in backcross generations. This is easier when the character is readily identified by visual inspection or simple test. Characters controlled by oligogenes are easiest to handle because they have a clear – cut phenotypic expression.

Table 9.1: Average Proportion of Genes from the Recurrent Parent in different Generations of a Backcross Programme

Generation	Average Proportion (in per cent) of Genes from the Recurrent Parent
F_1	50
BC_1	75
BC_2	87.5
BC_3	93.75
BC_4	96.875
BC_5	98.438
BC_6	99.218
BC_7	99.609
BC_8	99.805
BC_9	99.902
BC_{10}	99.951

Procedure of a Backcross Method for a Dominant Gene

Backcross method depends on whether the gene under transfer is dominant or recessive. Dominant genes have simpler plans than recessive because expression is both in heterozygous and homozygous forms.

Example: A high yielding and well adapted variety A is susceptible to stem rust. Variety B is resistant to stem rust. Resistance to stem rust is dominant over susceptibility (Figure 9.2).

Generalized Scheme

High yielding Resistant variety

After 5 to 6 backcrosses about 99 per cent of resistant genotype is retained.

Yr.7   Rust resistant plants are self-pollinated and seeds of selected plants are harvested separately. Selfing creates homozygosity.

Yr.8   Seeds from selected plants are grown separately and seeds from promising plants similar to A are bulked.

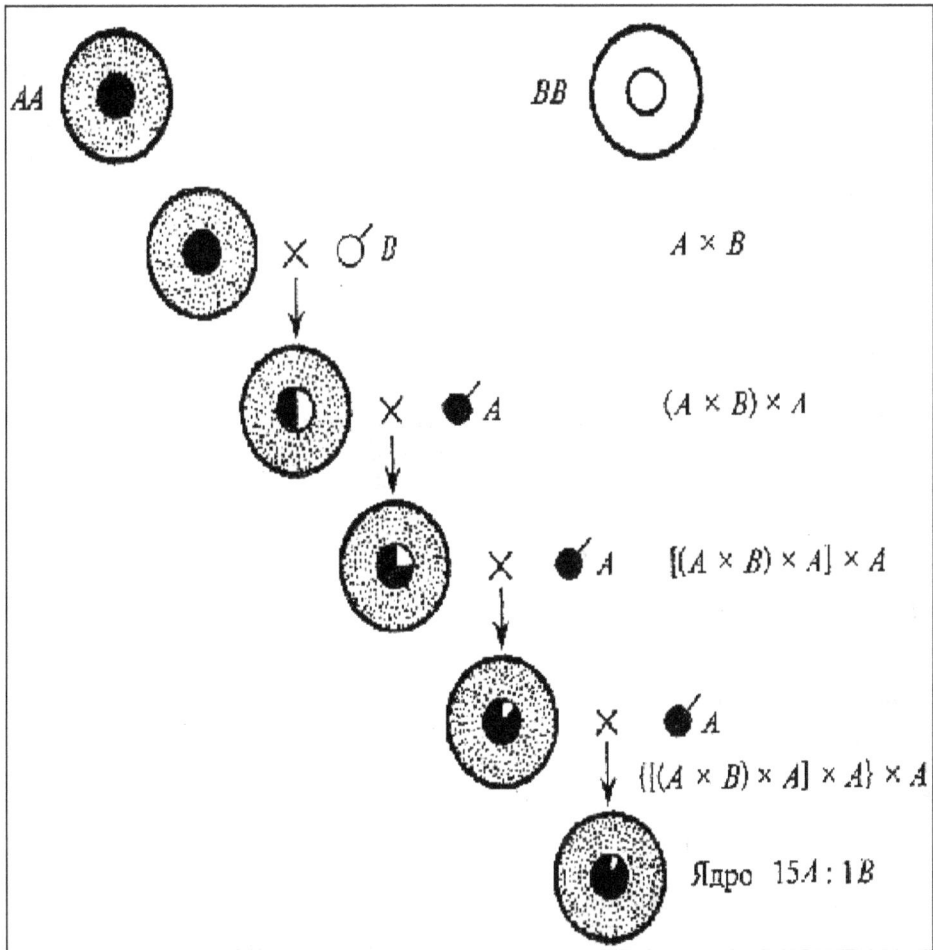

Figure 9.2: General Scheme of a Backcross.

Yr.9 The new variety is tested in a replicated yield trial with A as the standard.

Yr.10 If the new variety does better than A its seeds are multiplied and distributed to farmers.

Advantages of Backcross

☆ The improved variety is similar to the recurrent parent except for the new genes.

☆ Less time is spent in yield trial because its performance is known.

☆ Can be quickened by off season nurseries and green houses.

Disadvantages of Backcross

☆ The new variety is not superior to recurrent parent except for the genes being transferred.

☆ Constant backcrossing is difficult, time taking and costly.

☆ In normal conditions the improved variety takes a long time and may be overcome by other varieties.

9.3 Mass Selection

It was practiced in olden days by harvesting selected heads of plants, combining their seed to constitute planting material for the next generation. In this method it is assumed that the phenotype of selected plants reflects its genotypes. Selection is usually done on easily observable traits such as plant height, ear type, disease and insect resistance. Since several plants are selected and then seeds are composited, the selected population would be a mixture of similar looking purelines. Varieties developed through mass selection have a considerable genetic variation.

Objective

To improve local varieties especially in self-pollinated crops and to purify the existing pureline varieties.

Pureline – progeny of a single homozygous self-pollinated plant.

Procedure

A breeder can modify mass selection according to his needs.

Advantages of Mass Selection

☆ Varieties are more adapted than purelines because they result as a mixture of several purelines.

☆ Takes less time and cost because no crossing is involved.

☆ No need for extensive field trials.

Disadvantages of Mass Selection

☆ Mass selection does not generate variability but depends on existing variability in a population.

☆ It is difficult to know if the selected phenotype is due to genotype, environment or both.

☆ Plants being composited can be homozygotes or heterozygotes. Heterozygotes keep segregating calling for repeated phenotypic selection

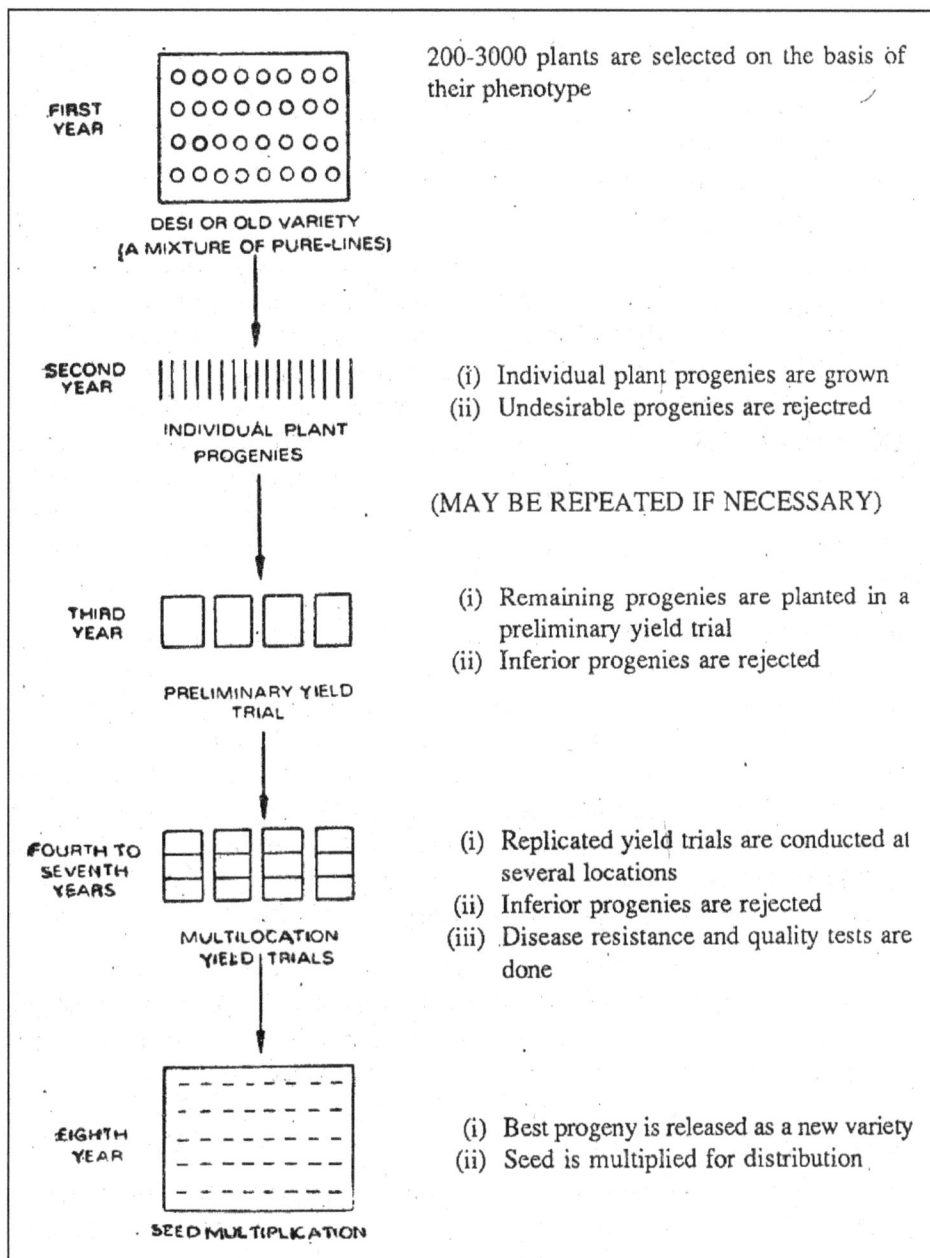

FIRST YEAR

DESI OR OLD VARIETY
(A MIXTURE OF PURE-LINES)

200-3000 plants are selected on the basis of their phenotype

SECOND YEAR

INDIVIDUAL PLANT PROGENIES

(i) Individual plant progenies are grown
(ii) Undesirable progenies are rejectred

(MAY BE REPEATED IF NECESSARY)

THIRD YEAR

PRELIMINARY YIELD TRIAL

(i) Remaining progenies are planted in a preliminary yield trial
(ii) Inferior progenies are rejected

FOURTH TO SEVENTH YEARS

MULTILOCATION YIELD TRIALS

(i) Replicated yield trials are conducted at several locations
(ii) Inferior progenies are rejected
(iii) Disease resistance and quality tests are done

EIGHTH YEAR

SEED MULTIPLICATION

(i) Best progeny is released as a new variety
(ii) Seed is multiplied for distribution

Figure 9.3: Generalized Scheme of Mass Selection.

9.4 Recurrent Selection

It is a cyclical selection that involves deriving desired lines from a mixed population by evaluating them, making all possible intercrosses among these selected lines and compositing the seeds before starting the next cycle.

The basic steps in the recurrent selection are:

☆ Identifying and self-pollinating superior plants.

☆ Testing of lines for desired character.

☆ Intercrossing the best selfed lines in all possible combinations.

☆ Compositing seeds from all intercrosses to produce new improved population in the next cycle.

The higher the frequency of desired gene combinations the greater the probability of getting better genotypes.

Procedure of a Simple Recurrent Selection

Simple recurrent selection is used for characters that are easy to observe phenotypically and have high heritability.

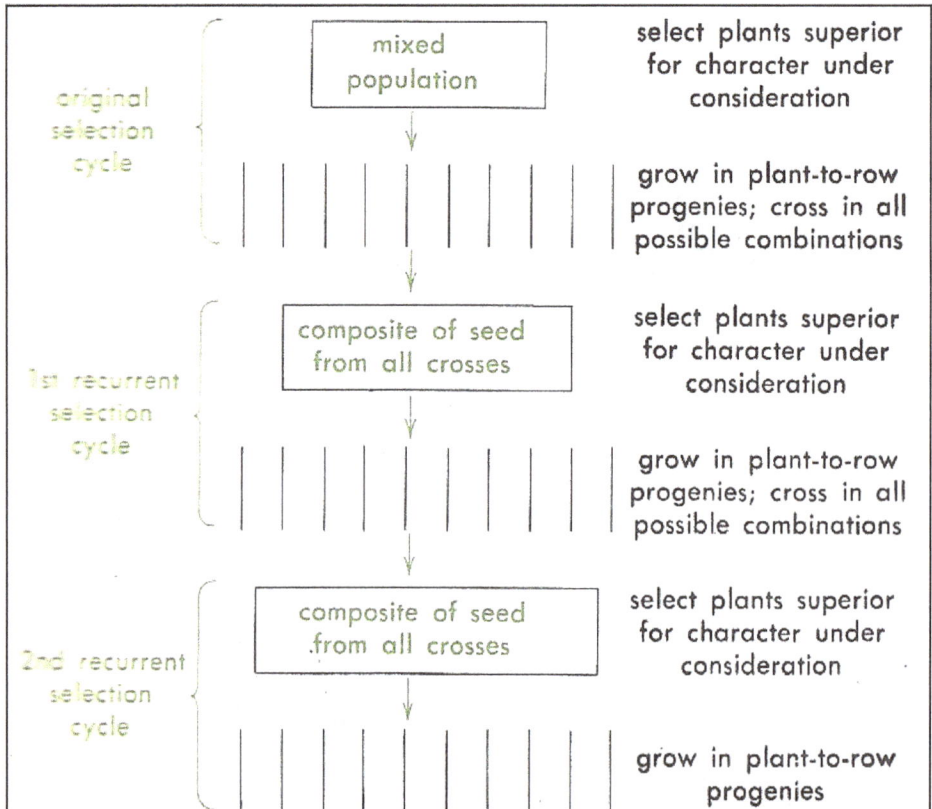

Figure 9.4: Generalized Scheme of Simple Recurrent Selection.

Note: In simple recurrent selection one cycle can be completed in only one season if the desired intercrosses between selected plants can be made in the same season. This has been useful in breeding for resistance against leaf rust whose symptoms appear before flowering.

Advantages of Recurrent Selection

☆ Highest performance is achieved from the least gene combinations in a population.

☆ More than one character can be improved in a population.

Disadvantages of Recurrent Selection

☆ It is expensive and time taking.

☆ It is limited to crops that don't experience self- sterility.

9.5 Pure-line Breeding

A pure line is the progeny of a single homozygous self-pollinated plant.

Note: There is no pure-line in cross pollinated plants.

Pure-line theory: It was established by a Danish Botanist, Johannsen in 1903.

He obtained purelines by selection of individual beans from a mixed lot. Within each of these pure lines he selected large and small seeds. Progenies of small seeds varied from progenies of large seeds, but the average weight of progenies from large seeds were quite similar to the average of progeny from small seeds within the same pureline. This indicated that *"selection within a mixed population of pure lines is effective in isolating lines that are different. Once the pure lines have been isolated, further selection within the line is ineffective"*.

In the original seed lot, variation in seed size was due to both genotype and environment but within the pure lines variation in seed size was due to environment only (Figure 9.5).

Characteristics of Pure Lines

☆ All the plants within a pureline have the same genotype as the plant from which the pure line was derived. This is because the parent plant was homozygous and self- pollinated.

☆ Variation within a pure line is environmental and non-heritable. Since all plants have identical genotype, selection within a pureline is not effective.

☆ Pure lines can be genetically variable due to mechanical mixture, natural hybridisation or mutation.

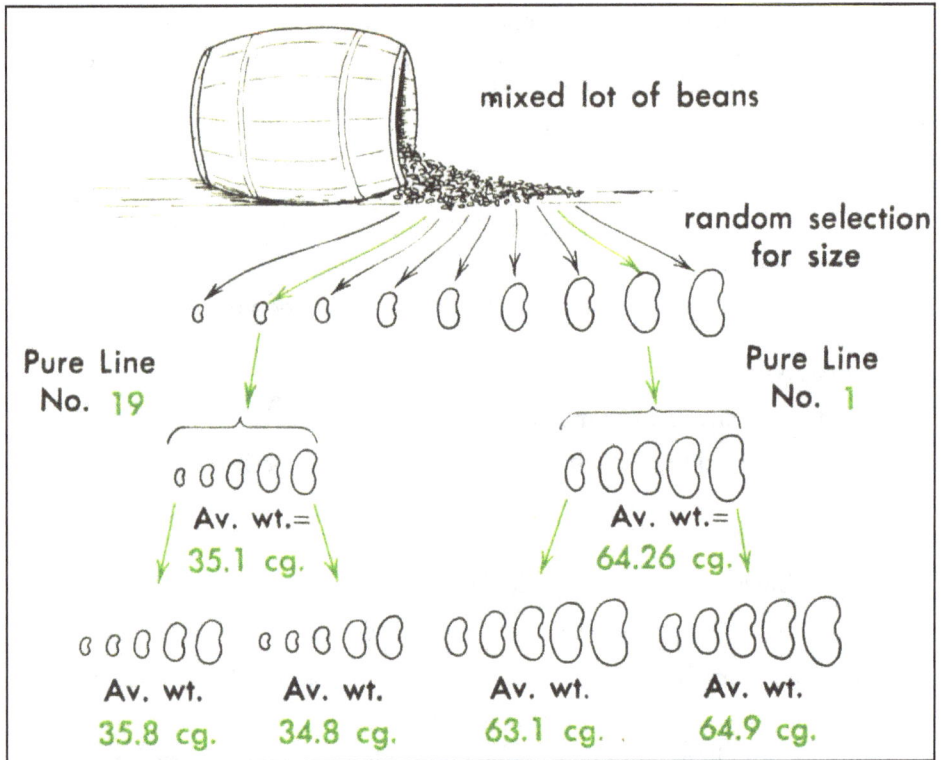

Figure 9.5: Pureline Selection as Proposed by Johannsen, 1903.

Objective

To improve self pollinated local varieties, old pure line varieties that have deteriorated and selection for new traits in a pure line.

Pure line selection follows 4 main steps:

1. Selection of individual plants from a local variety or a mixed population.
2. Visual evaluation of individual plant progenies.
3. Field trials.
4. Seed multiplication and distribution.

Advantages of Pureline Breeding

☆ Achieves maximum possible improvement over the original variety.

☆ Varieties are very inform and more liked by farmers.

☆ Seeds are easily identified during certification due to their uniformity.

FIRST YEAR

DESI OR OLD VARIETY
(A MIXTURE OF PURE-LINES)

200-3000 plants are selected on the basis of their phenotype

SECOND YEAR

INDIVIDUAL PLANT PROGENIES

(i) Individual plant progenies are grown
(ii) Undesirable progenies are rejectred

(MAY BE REPEATED IF NECESSARY)

THIRD YEAR

PRELIMINARY YIELD TRIAL

(i) Remaining progenies are planted in a preliminary yield trial
(ii) Inferior progenies are rejected

FOURTH TO SEVENTH YEARS

MULTILOCATION YIELD TRIALS

(i) Replicated yield trials are conducted at several locations
(ii) Inferior progenies are rejected
(iii) Disease resistance and quality tests are done

EIGHTH YEAR

SEED MULTIPLICATION

(i) Best progeny is released as a new variety
(ii) Seed is multiplied for distribution

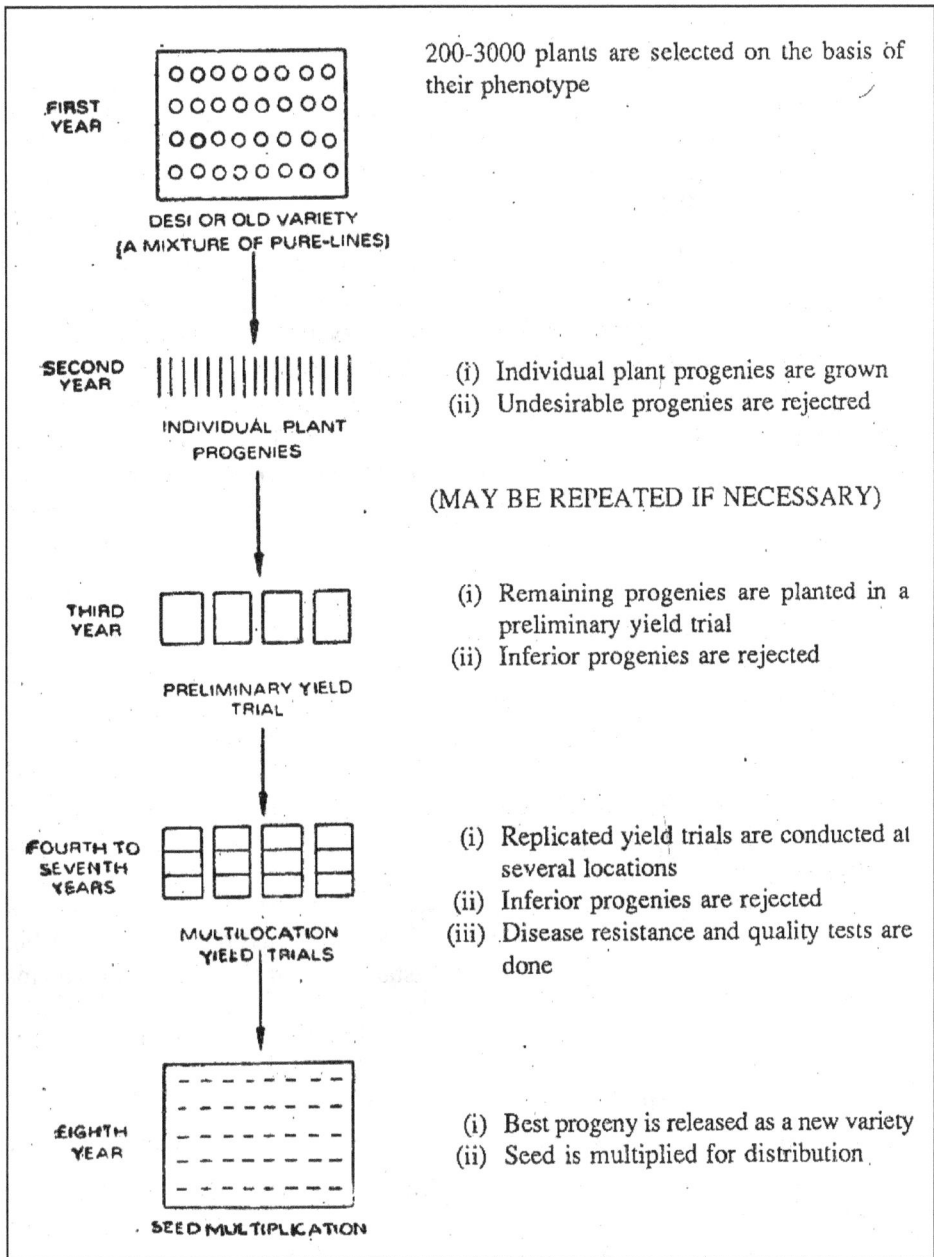

Figure 9.6: Generalized Scheme of Pureline Breeding.

Disadvantages Pureline Breeding

☆ Have less adaptation and stability than the original variety.

☆ Procedure needs more time, space and labour.

☆ Depends only on variation from the original variety.

☆ Applicable only to autogamous crops.

9.6 Clonal Selection

A clone is a group of plants produced from a single plant through asexual reproduction.

Some agricultural crops and a large number of horticultural crops are asexually propagated. Some examples are sugar-cane, potatoes, sweet potatoes, banana *etc.*

Characteristics of Clones

All individuals belonging to a single clone are identical in genotype because they are got only by mitotic cell division.

Phenotypic variation within a clone is due to environment only.

Phenotype of a clone is due to effects of genotype (G), environment (E) and interaction of G and E.

Clones can be maintained indefinitely from asexual reproduction. They degenerate due to viral or bacterial infections.

Genetic variation within clones may be due to somatic mutations, mechanical mixture, occasional sexual reproduction.

Procedure of Clonal Selection

Breeding of clones is different from sexually propagated crops because breeding material and commercial varieties are maintained by asexual reproduction. A big advantage of clonal selection is that a single outstanding plant in a population forms the basis of a new variety.

Breeding of clonal plants has 2 main phases:

1. Utilisation of variability or its creation by hybridization.

2. Selection of best genotype to produces a superior clone or variety.

Advantages of Clonal Selection

☆ Only method for selection in clonal crops.

☆ Avoids inbreeding depression.

☆ Purity of clones is maintained.

☆ A single outstanding plant can form a basis of developing a new variety.

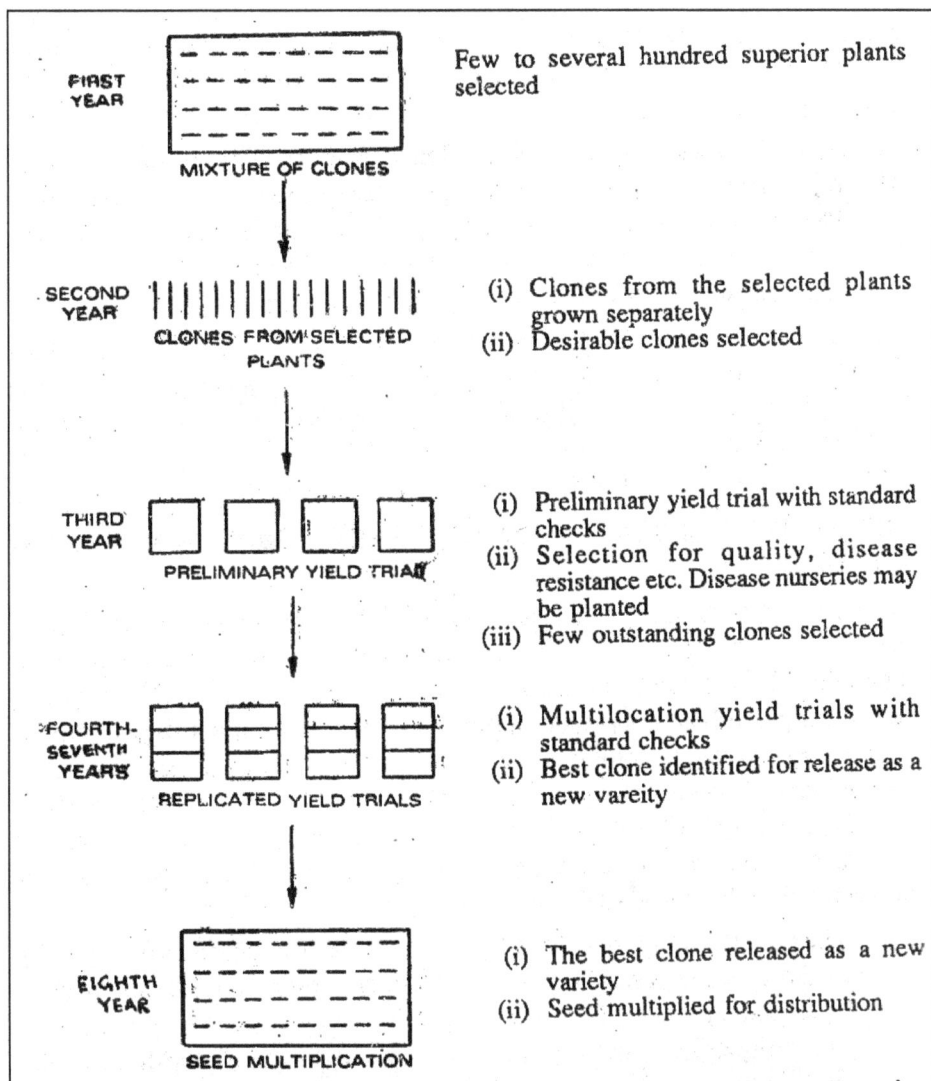

Figure 9.7: Scheme of Clonal Selection.

Disadvantages of Clonal Selection

- ☆ Undesired characteristics can easily be passed easily from parents to progenies.
- ☆ Handling of vegetative material is laborious.
- ☆ Reduced flowering and fertility. Some crops don't flower at all *e.g.* arrow roots. This makes it difficult for hybridization.

Summary

In this chapter we have looked at various breeding methods. We have also looked at their application, procedure and consequences. We have also seen factors to consider when choosing a plant breeding method. Advantages and disadvantages of each method have been discussed.

Self Test

1. Explain what you understand by the term reproduction coefficient and support your answer with 2 examples.

2. How can a breeder quicken a recurrent breeding method and to which crops can this be applied?

3. Describe the main objective of a conventional cross.

4. In a backcross method characters governed by oligogenes are preferred to those governed by polygenes. Explain why?

5. Why is selection of a recurrent parent a critical step in a backcross programme?

6. Discuss three differences and three similarities between clonal breeding and pureline breeding.

7. Describe clonal selection and its advantages.

Recommended Literature

1. B. P. Singh (1995). Plant Breeding. Kalyani Publishers, 677 pages.

2. V. L. Chopra (1989). Plant Breeding: Theory and Practice. Oxford Publishers. 471 pages.

3. A. Dafni. (2001). Field methods in pollination ecology. University of Haifa, Israel. P. 17-30.

4. J. M. Poehlman (1959). Breeding field crops. University of Missouri, 427 pages.

5. R. W. Allard (1960). Principles of plant breeding. John Wiley and Sons, Inc. California. 485 pages.

6. D. J. Van de have (1979). Heterosis in plant breeding, proceedings of the 7th Congress of EUCARPIA. Budapest. 365 pages.

Chapter 10

Release of New Varieties

10.0 Introduction

Enough seed must be grown to allow a new variety to be grown on a commercial scale in the areas where it is adapted. Varietal purity must be maintained.

10.1 Classes of Seeds

Three classes of pure seed are recognized by the International Crop Improvement Association, (ICIA).

1. Breeder seed – (White label) it is seed or vegetative material produced directly by the breeder.

2. Foundation seed – (Blue label) it is seed increased from breeder seed.

3. Certified seed – (Red label) it is seed grown by certified seed growers on large scales.

Before a variety can be certified, it must be approved by a Government Board *e.g.* (KEPHIS) in Kenya.

Certification procedure follows this general pattern:

☆ Seed grower plants foundation, registered or certified seed of an approved variety.

☆ Ground must not have been planted the same crop or related species for a specified time. This ensures genetic purity.

☆ Minimum isolation distances must be observed in cross pollinated species. *e.g.* sunflower 1000m, wheat 200m, maize 500m. In presence of a barrier this distance is less.

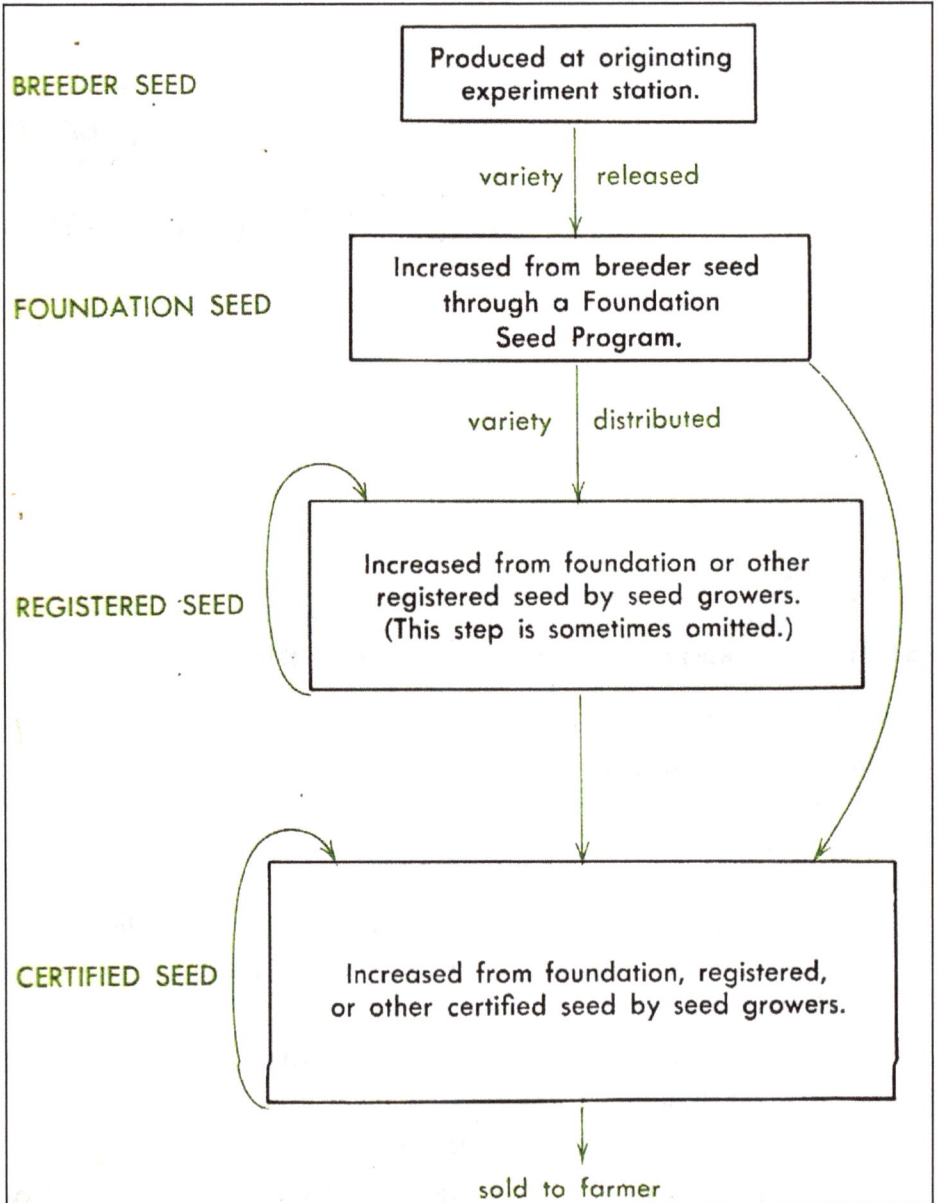

BREEDER SEED

Produced at originating experiment station.

variety | released

FOUNDATION SEED

Increased from breeder seed through a Foundation Seed Program.

variety | distributed

REGISTERED SEED

Increased from foundation or other registered seed by seed growers. (This step is sometimes omitted.)

CERTIFIED SEED

Increased from foundation, registered, or other certified seed by seed growers.

sold to farmer

Figure 10.1: General Scheme of Production of a Variety.

10.2 Plant Breeders' Rights

This was established in Kenya in 1998 to patent breeders' work.

Before a variety can be patented it must undergo a Distinct, Uniformity and Stability (DUS) test.

Distinctive – difference from other related varieties in shape, size, maturity, appearances *etc.*

Distinction can be done by molecular methods such as electrophoresis and RAPDs. It is tested in two growing seasons.

1. Uniformity – sameness of a crop in a field. It is best expressed in vegetatively propagated species.

2. Stability – consistence in maintaining essential characteristics in differing locations and generations

A variety must undergo National Performance Trial (NPT). This tests the value of the variety for cultivation in all potential parts of the country.

Summary

In this chapter we have discussed the procedure of releasing new varieties in Kenya. We have also looked at various classes of seeds. Plant breeders' rights have also been discussed.

Self Test

1. Define DUS test.

2. What is the role of ISTA in plant breeding?

3. Outline the main steps towards seed certification.

4. Describe the role of KEPHIS in seed industry in Kenya.

Recommended Literature

1. A. Dafni. (2001). Field methods in pollination ecology. University of Haifa, Israel. pp. 17-30.

2. J. M. Poehlman (1959). Breeding Field crops. University of Missouri, 427 pages.

3. R. W. Allard (1960). Principles of plant breeding. John Wiley and Sons, Inc. California. 485 pages.

4. D. J. Van de have (1979). Heterosis in plant breeding, proceedings of the 7th Congress of EUCARPIA. Budapest. 365 pages.

Samples of Past Examination Questions and Tests

EXAM SAMPLE 1

Answer any TWO questions in Section A and All questions in Section B.

Answers in Section B should be written in the spaces provided.

SECTION A: Essay Questions (20 Marks Each)

1. Discuss the main features in each of the following breeding methods:
   a) Backcross
   b) Clonal selection
   c) Pure-line breeding

2. Write an essay on crop genetic resources.

3. Discuss the various modes of reproduction in crop plants and their significance in plant breeding.

4. Discuss genetics of disease and pest resistance in crop plants.

SECTION B: Short Answer Questions (6 Marks each)

1. Outline three main differences between breeding self pollinated and cross-pollinated crops.

2. Explain the main steps in mass selection.

3. A plant breeder is studying genetics of bean resistance to Anthracnose *Collectotrichum lindemuthianum*. He crossed susceptible with resistant varieties and observed segregation pattern as shown below:

Genotype	Populations	Total Seeds	Observed Number	
			Resistant	Susceptible
A	P_1	50	50	0
B	P_2	43	0	43
AXB	F_1	40	0	40
AXB	F_2	296	16	280

    a. What mode of inheritance did he observe?

    b. State the features associated with this type of inheritance.

4. Discuss six methods of suppressing self-incompatibility in crop plants.

5. Explain, giving examples the meaning of the following terms:
    a) Microcentre
    b) Apomixis
    c) Topcross
    d) Macromutation
    e) Gene for gene relationship
    f) Pathogenecity

EXAM SAMPLE 2

Answer All Questions

1. Define the following terms as applied in plant breeding: *(4 marks each)*
    a) Pseudoresistance
    b) Gene mutations
    c) Tissue culture in resistance breeding
    d) Pure line breeding

2. Discuss the main features of the following methods of breeding:
    a) Recurrent selection (8mks)
    b) Mass selection (7mks)

3. Seed bruchids *Zabrotes subfasciatus* are a major pest of beans *Phaseolus vulgaris*. Resistance to this pest is inherited as a single dominant gene. Describe a method that could be used for incorporating resistance to this pest from the donor to the recurrent parent.

4. A plant breeders is studying the genetics of potato resistance to late blight *caused by Phytophthora infestans.* She made crosses among three cultivars (two resistant – TP – 1084 and TP – 1180 and one susceptible TP-0984) and grew F_1 and F_2 progenies. Segregation for resistance to late blight was as follows:

Cultivars/Cross	Generation	Total Plants	Observed Resistant	Number Susceptible
TP – 1084 (A)	F_1	30	30	0
TP-1180 (B)	P_2	31	31	0
TP – 0984 (S)	P_3	30	0	30
A X S	F_1	35	35	0
B X S	F_1	38	38	0
A X B	F_2	191	179	12

Use the data above to determine:

(a) Mode of inheritance of potato resistance to late blight.

(b) Allelic relationships among resistance genes present in cultivars TP – 1084 and TP – 1180 (8 marks).

EXAM SAMPLE 3

Answer All Questions

1. Discuss the following:
   a) Vertical resistance
   b) Pseudo resistance
   c) Pure line breeding
   d) Active resistance

2. In an attempt to combat pests and diseases in crops, scientists and farmers may instead predispose crops to attack. Giving specific examples, discuss the role played by plant breeders and farmers in predisposing crops to attack.

3. Discuss the main features of the following methods of breeding:

a) Recurrent selection

b) Bulk population breeding

c) Backcross breeding

4. A plant breeder is studying the genetics of bean resistance to seed Bruchids. He crossed a resistant and a susceptible line and grew F_1 and F_2 generations. Segregation for resistance to Bruchids was as follows:

Genotype	Population	Total Seeds	Observed Resistant	Observed Susceptible
A	P_1	50	50	0
B	P_2	43	0	43
AXB	F_1	40	0	40
(AXB) (AXB)	F_2	296	16	280

Use data in the above table to determine:

a) Mode of inheritance of bean resistance to Bruchids

b) If F_3 progenies were raised from selected susceptible F_2 plants, what kind of segregation would you expect?

EXAM SAMPLE 4

Answer and TWO questions in Section A and ALL questions in Section B.

Answers in Section B should be written in the spaces provided on the question paper.

SECTION A: Essay Questions (20 Marks each)

1. Mwitemania is a local high yielding bean variety that has become susceptible to Anthracnose *Collectotrichum lindemuthianum*. Discuss a suitable method that could be used for incorporating resistance to Anthracnose from the donor to the recurrent parent.

2. With the help of suitable examples, discuss evolution of cultivated crop plants.

3. Discuss the main features in each of the following breeding methods:

a) Mass selection

b) Clonal selection

4. A plant breeder is studying genetics of potato resistance to late blight caused by *Phytophora infectans*. She made crosses among three cultivars (two resistant KU – 100 and KU-200 and one susceptible JK – 50). She grew F_2 and F_3 progenies and observed segregation for resistance to late blight as follows:

Cultivars/Cross	Generation	Total Plants	Observed Numbers	
			Resistant	Susceptible
KU - 100	P_1	40	40	0
KU –200	P_2	41	41	0
KU - 50	P_3	40	0	40
KU –100 X JK – 50	F_1	35	35	0
	F_2	280	215	65
KU–200 X JK - 50	F_1	38	38	0
	F_2	189	140	49
KU-100 X KU-200	F_1	33	33	0
	F_2	191	179	12

Use the data above to determine:

a) Mode of inheritance of resistance to late blight.

b) Allelic relationships among resistance genes present in cultivars KU-100 and KU-200.

SECTION B: Short Answer Questions (6 Marks each)

1. A wheat breeder needs genes for drought tolerance. Give six types of genetic resources that would serve this purpose.

2. What techniques are used for suppressing self- incompatibility in crops?

3. Briefly discuss morphological and physiological factors of insect resistance in crops.

4. Explain six mechanisms that encourage allogamy.

5. Botany students want to start a breeding programme in the Department. What factors must they consider?

EXAM SAMPLE 5

ANSWER ALL questions in SECTION A and ONE question in SECTION B.

Answers in SECTION A should be written in the spaces provided on the question paper.

SECTION A: Short Answer Questions (5 marks each)

1. Describe five techniques of suppressing self-incompatibility in crops.

2. Outline factors to consider when choosing a plant breeding method.

3. Explain, giving examples, the meaning of the following terms in plant breeding:
   a) Top cross
   b) Biotype
   c) Pure line
   d) Gene for gene relationship
   e) NPT

4. Explain five mechanisms that encourage autogamy.

5. Discuss three patterns of evolution of cultivated crops.

6. Give five differences between oligogenic and polygenic modes of inheritance.

7. A plant breeder needs genes for disease resistance. Explain five types of genetic resources you would recommend.

8. Briefly discuss the procedure of releasing new varieties.

SECTION B: Essay Questions (30 Marks)
Answer any ONE question

1. Discuss the following:
   a) Techniques in breeding field crops.
   b) Pest and disease resistance in crops.

2. Write a comprehensive essay on:
   a) Male sterility in plants and methods of overcoming it.
   b) Evolution of cultivated crop plants.

3. Discuss the main features in each of the following breeding methods:

   a) Recurrent selection

   b) Clonal selection

   c) Backcross

   d) Mass selection

EXAM SAMPLE 6

Answer ALL questions in Section A and ONE question in Section B.

Answers in Section A should be written in the spaces provided on the question paper.

SECTION A: Essay Questions (5 Marks Each)

1. Discuss characteristics of clones.

2. Describe five techniques of suppressing self-incompatibility in crops.

3. Outline five main differences between breeding self pollinated and cross pollinated crops.

4. Briefly discuss male sterility and its occurrence in crops.

5. Explain giving examples the meaning of the following terms in plant breeding:

   a) Diallel cross

   b) Apogamy

   c) Gene for gene relationship

   d) Biotype

   e) Tester

6. Explain five differences between oligogenic and polygenic modes of inheritance.

7. Discuss five major factors to consider when choosing a plant breeding method.

8. Explain five mechanisms that encourage autogamy.

SECTION B: Essay Questions (30 Marks)
Answer any ONE question

1. Discuss various types of crosses and their application in plant breeding.

2(a). A plant breeder is studying the genetics of potato resistance to late blight caused by *Phytophthora infestans*. She made crosses among three cultivars (Two resistant TP – 1084 and one susceptible TP – 0984) which gave F_1 and F_2 progenies. Segregation for resistance to late blight was as follows:

Cultivars/Cross	Generation	Total Plants	Observed Numbers	
			Resistant	Susceptible
TP – 1084 (A)	P_1	40	40	0
TP – 1180 (B)	P_2	41	41	0
TP – 0984 (S)	P_3	40	0	40
	F_1	35	35	0
A X S	F_2	399	301	98
	F_1	38	38	0
B X S	F_2	189	140	49
	F_1	33	33	0
A X B	F_2	254	238	16

Use the data above to determine:

a) The mode of inheritance of potato resistance to late blight.

b) Allelic relationship among resistance genes present in cultivars TP – 1084 and TP - 1180

2(b). Discuss patterns of evolution of cultivated crop.

3. Write a comprehensive essay on the world centers of genetic diversity.

EXAM SAMPLE 7

*Answer **ALL** questions in **Section A** and **ONE** question in **Section B**.*

*Answers in **Section A** should be written in the **spaces provided** on the question paper.*

SECTION A: Essay Questions (5 Marks Each)

1. Discuss three patterns of evolution of cultivated crops.

2. Briefly discuss the procedure of releasing new varieties.

3. List five main differences between breeding self pollinated and cross pollinated crops.

4. Briefly discuss morphological and physiological factors of disease resistance in crops.

5. Explain giving examples the meaning of the following terms in plant breeding:
   a) Microcenter
   b) Apogamy
   c) Gene for gene relationship
   d) Pure line
   e) CMS

6. Discuss differences between oligogenic and polygenic modes of inheritance.

7. Discuss five types of genetic resources.

8. Explain five mechanisms that encourage allogamy.

SECTION B: Essay Questions (30 MARKS)
Answer any ONE question

1. Discuss five common plant breeding methods.

2. Write an essay on the main achievements and challenges in plant breeding today.

3. Discuss the following:
   a) Techniques in breeding field crops.
   b) Male sterility in plants and methods of overcoming it.

EXAM SAMPLE 8

Answer ALL questions in Section A and ONE question in Section B.

Answers in Section A should be written in the spaces provided on the question paper.

SECTION A: Essay Questions (5 Marks Each)

1. Discuss characteristics of pure lines.

2. Briefly discuss the procedure of releasing new varieties.

3. Differentiate between breeding self pollinated from cross pollinated crops.

4. Briefly discuss morphological and physiological factors of insect resistance in crops.

5. Explain giving examples the meaning of the following terms in plant breeding:

   a) Diallel cross

   b) Parthenogenesis

   c) Gene for gene relationship

   d) Tester

   e) CMS

6. Compare oligogenic with polygenic modes of inheritance.

7. Discuss five major factors to consider when choosing a plant breeding method.

8. Explain five mechanisms that encourage allogamy.

SECTION B: Essay Questions (30 MARKS)
Answer any ONE question

1. Discuss five common plant breeding methods.

2. Write an essay on the main achievements and challenges in plant breeding today.

3. Discuss the following:

   a) Techniques in breeding field crops.

   b) Self incompatibility in plants and methods of overcoming it.

Index

www.ingramcontent.com/pod-product-compliance
Lightning Source LLC
Chambersburg PA
CBHW050519190326
41458CB00005B/1589